ADVANCED III-V SEMICONDUCTOR MATERIALS TECHNOLOGY ASSESSMENT

Edited by

M. Nowogrodzki

RCA Laboratories
Princeton, New Jersey

np

NOYES PUBLICATIONS
Park Ridge, New Jersey, USA

Library of Congress Catalog Card Number: 83-22132
ISBN: 0-8155-0974-X
ISSN: 0198-6880
Printed in the United States

Published in the United States of America by
Noyes Publications
Mill Road, Park Ridge, New Jersey 07656

Library of Congress Cataloging in Publication Data

Main entry under title:

Advanced III-V semiconductor materials technology
assessment.

(Chemical technology review ; no. 225)
Bibliography: p.
Includes index.
1. Microwave devices. 2. Transistors, Microwave.
3. Gallium arsenide. I. Nowogrodzki, M. II. Title:
Advanced 3-5 semiconductor materials technology assessment. III. Title: Advanced three-five semiconductor
materials technology assessment. IV. Series.
TK7876.A3 1984 621.381'33 83-22132
ISBN 0-8155-0974-X

Foreword

This book presents an extensive survey of the state of the art in the field of III-V semiconductor materials and devices for operation at microwave frequencies. Competing technologies and physical device limitations are discussed, as are projected difficulties in implementing emerging technology. Requirements for future space communications systems are identified, and specific R&D programs for future systems components are recommended.

Current status and trends in materials technology are examined with emphasis on bulk growth of semi-insulating GaAs and InP, epitaxial growth, and ion implantation. Microwave solid-state discrete active devices and multigigabit rate GaAs digital integrated circuits are covered; and the development of GaInAs devices, heterojunction devices, and quasi-ballistic devices are considered. Fundamental limits of semiconductor devices and problems in implementation are also included.

Recommendations for technology exploration and advancement, with alternative approaches, include—in the device area—the need to establish limits in the frequency and linearity of three-terminal devices, and—in the digital area—the need for development of enhancement-mode logic circuits.

The information in the book is from *Advanced III-V Semiconductor Technology Assessment,* prepared by M. Nowogrodzki of RCA Laboratories, for the National Aeronautics and Space Administration Lewis Research Center, February 1983.

The table of contents is organized in such a way as to serve as a subject index and provides easy access to the information contained in the book.

Advanced composition and production methods developed by Noyes Data are employed to bring this durably bound book to you in a minimum of time. Special techniques are used to close the gap between "manuscript" and "completed book." In order to keep the price of the book to a reasonable level, it has been partially reproduced by photo-offset directly from the original report and the cost saving passed on to the reader. Due to this method of publishing, certain portions of the book may be less legible than desired.

Acknowledgments

This book describes the results of a study performed at the Microwave Technology Center of RCA Laboratories for NASA Lewis Research Center. The director of the Microwave Technology Center is Dr. Fred Sterzer. The project coordinator was M. Nowogrodzki. The individual tasks of the study were assigned to different individuals, who also wrote the draft reports for these tasks, as follows:

Task 1: Identify space communications applications–M. Nowogrodzki

Task 2: Assess present status of technology–Dr. S.Y. Narayan

Task 3: Identify domains of competing technologies–Dr. E.F. Belohoubek

Task 4: Identify fundamental physical limits–Dr. H.C. Huang

Task 5: Identify problems in implementation–P.D. Gardner, M. Nowogrodzki, Dr. W.J. Slusark, Jr., Dr. L.C. Upadhyayula, H. Wolkstein

Task 6: Develop recommendations for implementation–M. Nowogrodzki

NOTICE

List of Abbreviations

BFL	buffered FET logic
BJT	homojunction bipolar transistor
BW	bandwidth
CMOS	complementary metal oxide semiconductor
CVD	chemical-vapor deposition
DBS	Direct Broadcast Satellite
DCFL	directly coupled FET logic
DLTS	deep-level trap spectroscopy
ECL	emitter-coupled logic
EM	electromigration
FET	field-effect transistor
d-FET	depletion-mode FET
e-FET	enhanced-mode FET
FI	fan-in
FIT	failure in test
FO	fan-out
HBT	heterojunction bipolar transistor
HEMT	high-electron mobility transistor
HJFET	Heterojunction Gate GaAs FET
HMIC	hybrid microwave integrated circuit
IC	integrated circuit

ILO	injection-locked oscillators
IMD	intermodulation distortion
IMPATT	impact avalanche transit-time
JJ	Josephson junction
LEC	liquid-encapsulated Czochralski
LED	light-emitting diode
LPE	liquid-phase epitaxy
LPFL	low pinchoff FET logic
LSI	large scale integration
MBC	miniature beryllia circuit
MBE	molecular-beam epitoxy
MESFET	metal-semiconductor field effect transistor
d-MESFET	depletion mode MESFET
e-MESFET	enhancement mode MESFET
MIC	microwave integrated circuit
MISFET	metal-insulator-semiconductor field effect transistor
MMIC	monolithic microwave integrated circuit
MOCVD	metal/organic chemical-vapor deposition
MOS	metal oxide semiconductor
MSI	medium scale integration
MTBF	mean time between failures
PBN	pyrolitic boron nitride
PBT	permeable-base transistor
PDB	planar-doped barrier
PRBS	pseudorandom bit sequence
Q-A	quadrature-amplitude
RHEED	reflection high-energy electron diffraction
RO	ring oscillator
SAINT	Self-aligned Implantation for n^+-Layer Technology
SDFL	Schottky-diode FET logic
SI	semi-insulating
SIMS	secondary ion mass spectrometry
SMART	Solar Microwave Array Technology
SSB	single-sideband

SSPA	solid-state power amplifiers
TED	transferred-electron devices
TEG	triethylgallium
TEGFET	two-dimensional electron gas field-effect transistor
TMG	trimethylgallium
TWT	traveling-wave tube
TWTA	traveling-wave tube amplifier
UHV	ultrahigh vacuum
VHSIC	very high speed integrated circuit
VLSI	very large scale integration
VPE	vapor-phase epitaxy

Contents and Subject Index

Executive Summary

Under Contract NAS-3-22884, the impact of the state of the art in the area of III-V semiconductor materials employed in microwave solid-state components on future space communication systems was studied. Following the program format outlined by NASA, the study was pursued in terms of distinct tasks, as follows:

Task 1: Space Communications Applications

Task 2: Present Status of Technology

Task 3: Competing Technologies

Task 4: Fundamental Limits

Task 5: Problems in Implementation

Task 6: Recommendations for Implementation

In general, the investigations followed the plan developed in the original proposal. The two deviations from that plan concerned Tasks 1 and 2.

In Task 1, it was originally envisaged that future space communication systems could be identified on the basis of existing planning documents in sufficient detail to make possible the development of broad component "specifications," which would then serve as the background for Tasks 3, 4, and 5. This did not prove feasible, since few plans offering sufficiently precise projections - in terms of frequency, concept, or system architecture - could be gleaned from the literature. Thus, instead of very specific component specifications, the needs identified in Task 1 were broader in scope, based on subjective projections of what technical innovations future systems might contain, either in terms of evolutionary progress from present thinking (e.g., "large structures") or as a result of basic changes in technical approach (e.g., "reconfigurable transponders").

Although originally planned as one of a number of "coequal" tasks, it became apparent that Task 2 (Present Status of Technology) could well serve as the single most important task, central to the study. We therefore allowed (with the concurrence of the contract monitor) the expansion of this effort in terms of both scope and allocated time.

TASK 1: SPACE COMMUNICATIONS APPLICATIONS

The original intent was to define future space communications applications in detail sufficient to make possible the development of component specifications

- in terms of such parameters as frequency, function, mechanical characteristics, and reliability - to serve as a basis for the study. This did not prove feasible because of the lack of existing planning information on which such specifications could be based. It would, of course, have been possible to invent "future systems" solely for the purpose of defining the components they might require. Based on the program objectives, as discussed during an early planning meeting with NASA/LRC personnel (in September 1981), namely, the preparation of recommendations for NASA sponsorship of new broad technology development programs (rather than those for specific components) for future space communication systems, it was judged more productive to attempt to project future systems in terms of their more general characteristics and to extract new technological needs for systems so defined.

As a result of literature searches, meetings with personnel from technical and operating activities both within and outside RCA, and personal interviews with various individuals, the projections of future systems fell into three main categories:

(1) Systems that were logical extensions of systems currently being planned for the 1985-1995 time slot.

(2) Systems in which new types of modulation schemes would be employed.

(3) Novel system concepts that may strongly impact future space communications.

Evolutionary systems concepts considered in the study were (a) large antenna systems; (b) satellite crosslinks; (c) single-sideband systems; and (d) digital communication systems. Spread-spectrum techniques and quadrature-amplitude modulation are examples of advanced modulation approaches that might be used in future systems. Transponders that can be reconfigured by Earth command to change their electrical characteristics (e.g., from linear multi-carrier to saturated operation) and a concept originally proposed for the Solar Power Satellite involving a distributed antenna in which solar cells are placed on one side of a "space blanket" so that they can provide dc power directly to the antenna elements placed on the other side were considered as examples of novel concepts for future systems.

TASK 2: PRESENT STATUS OF TECHNOLOGY

In this section of the report, emphasis has been placed on materials and discrete device technology since this forms the basis of microwave and digital integrated circuits. We have attempted to project future trends and have included a subsection on the exploratory development that is being undertaken in many laboratories.

The field of III-V compound microwave and gigabit-rate logic technology is currently very active. We have tried to include most of the results published up to early 1982. The trends discussed in this section are believed to be reasonably accurate. The extensive literature references provided are representative rather than complete or in historical order.

We believe that a major change is occurring in III-V compound technology. The major attraction of GaAs for microwave and gigabit-rate digital ICs until recently was due mainly to the superior semiconducting properties of GaAs and not to its fabricating technology. New materials growth techniques such as molecular-beam epitaxy (MBE) and metal/organic chemical vapor deposition (MOCVD) now permit an unprecedented complexity, flexibility, and diversity in III-V compound growth. This technological strength, in addition to superior fundamental semiconducting properties, promises to make III-V compound research an exciting and useful endeavor in the mid-1980 to 2000 time frame.

The major topics reviewed in this section are:

1. Materials technology, including bulk growth of substrates, epitaxial crystal growth, and ion-implantation doping techniques.
2. Microwave solid-state discrete active devices, including two-terminal and three-terminal devices.
3. Multigigabit-rate GaAs digital integrated circuits.
4. Microwave integrated circuits, including both hybrid and monolithic circuits.
5. Exploratory developments currently being pursued, including GaInAs devices, heterojunction devices, and quasi-ballistic devices.

TASK 3: COMPETING TECHNOLOGIES

The main topics discussed in this section of the report are (a) rf power generation; (b) filter structures; and (c) microwave circuit fabrication. Each of these will profoundly influence the communication systems of the future.

In the area of power generation, the present trend of replacing thermionic devices with solid-state amplifiers in the spacecraft transponders is likely to continue, certainly at the lower frequencies. Various types of solid-state sources (FETs vs IMPATTs, GaAs vs Si) are likely to be used for transponders operating at different frequencies and power levels. Active antenna arrays, in which large numbers of antenna elements whose power is combined (in space) for fixed-, switched-, or scanning-beam systems are a very probable means of future generation of large amounts of power.

In the filter area, miniaturized active filters and new materials employed in dielectric resonators are expected to replace the present bulky and relatively heavy channel filters.

There are three distinct ways in which microwave circuits can be fabricated: In the conventional hybrid technology, active devices (packaged or in chip form) are mounted in custom-made microwave matching circuits, employing either distributed (transmission-line segments) or lumped-constant (capacitors, inductors) elements. In monolithic circuits, both active and passive circuit components are fabricated at the same time. In an emerging technology employing miniature hybrid circuits, the passive circuits are batch-fabricated and therefore quite cost effective, while the active elements (e.g., FETs) are later placed in the circuits, possibly employing automated bonding methods. This latter technology may prove superior in systems where fully monolithic approaches may not be justified, for either technical or economic reasons.

TASK 4: FUNDAMENTAL LIMITS

The solid-state devices discussed in this section of the report can be functionally characterized to fit into two categories: linear devices and logic devices. Linear devices include both low-noise and high-power devices for transponder applications; logic devices are mostly for baseband signal processing.

For future applications, the transponder carrier frequency is likely to be higher than it is now - 20 GHz and above - because of the crowded lower-frequency spectrum. On the other hand, the trend of baseband signal processing may be expected to head toward complex, multifunctional processing to facilitate on-board switching functions.

In this task, we concentrated our study on the fundamental limits in high-frequency microwave monolithic integrated-circuit (MMIC) devices and in high-density logic devices. The fundamental physical limits such as electrical, thermal, and physical-dimension limitations are presented. Research areas in new device concepts as well as technology development that will affect the limits are also identified.

TASK 5: PROBLEMS IN IMPLEMENTATION

The recently completed major effort by RCA to develop, space-qualify, and transfer to a manufacturing operation an all-solid-state transponder for use in a commercial communication system provided valuable insights into the problem of implementing a new technology for a critical operation. These experiences - in terms of both manufacturing and test approaches and reliability-proof considerations - are reviewed in this section of the report. In addition, a new type of solid-state amplifier, the GaInAs MISFET, which may well provide linearity properties superior to those in present-day GaAs units, is discussed, as are the existing technology limitations in realizing high-speed digital monolithic circuits.

TASK 6: RECOMMENDATIONS FOR IMPLEMENTATION

This section contains recommendations on how to take advantage of the technical capabilities of III-V compounds so as to enhance the performance of future space communication systems. The following is noted:

(a) The recommendations are for technology-development programs, as opposed to specific mission-related programs. The RCA SSPA development for the SATCOM system may serve as an illustration: The technology of GaAs power FETs operating at 4 GHz was considered "ready" for exploitation several years ago. It took considerable skill, tight planning, a sizable commitment of funds, and several important ancillary programs - such as reliability proof, automated testing, "transfer of technology" to a manufacturing group, and the establishment of quality-assurance procedures - before this "ripe" technology could be utilized in spaceborne transponders.

(b) In estimating costs, it was assumed that most of the technology development would be done under contract. Contract administration and technical monitoring costs are NOT included in the projections, which are in 1982 dollars.

(c) III-V laser components are not included in the study. Also, programs that have been judged adequately covered by present NASA-sponsored effort were not included in the recommendations. As an example, ongoing and presently planned effort in the area of switches and switch matrices will likely result in components adequate for future space systems.

To focus more precisely on the recommended technology-development programs, classes of applications and their component needs were examined and tabulated. The results are presented in table VII-1 of the report.

The following is a list of programs recommended for implementation:

1(a) Antenna Module Technology (Monolithic)
(b) Antenna Module Technology (Miniature Hybrid)
(c) Antenna Module Technology (SMART)
2(a) Adjustable Transponder Components (Switched)
(b) Adjustable Transponder Components (Electric)
3 Ternary-Compound Linear Amplifiers
4 High-Speed Digital Circuit Technology
5(a) Millimeter-Wave Device Investigations (Ternary Compounds)
(b) Millimeter-Wave Device Investigations (Permeable-Base Transistor)
(c) Millimeter-Wave Device Investigations (Quasi-Ballistic)
6 High-Efficiency, Lightweight, Miniature Ground Transponder

CONCLUDING REMARKS

Future space communication systems are likely to show technical advances that, for purposes of analysis, can be grouped into three categories: (1) extensions of present systems; (2) systems in which new types of modulation will be employed; and (3) novel system concepts based on progress in components technologies. The study described in this report examined microwave components based on III-V compounds that require development to fill these future system requirements.

Large antenna structures employed in future systems will require quantities of antenna modules which - depending on the frequency of operation and antenna characteristics - could be fabricated in either monolithic or miniature-hybrid form. To achieve better utilization of the crowded spectrum, "superlinear" amplifiers will be likely to replace units in the currently used frequency region. The concept of a transponder with characteristics (bandwidth, linearity) changeable on command from Earth offers a flexibility not obtainable with present-day systems. Intersatellite links of the future may well operate at millimeter wavelengths, possibly doing away with primary power distribution problems by the use of solar cells to power amplifier modules directly. Digital systems will require high-speed monolithic ICs.

Advances in III-V materials technology (molecular-beam epitaxy, organometallic epitaxy, the use of vapor-phase epitaxy for growing ternary and quaternary compounds) have made new transistor geometries possible. These show promise of outperforming conventional planar GaAs FETs at microwave and millimeter-wave frequencies. This may be a particularly fruitful area of research.

The conclusions of the study are contained in a number of specific recommendations for research programs aimed at components to be employed in the identified applications.

I

Introduction

This report presents the findings of a 12-month study of the state of technology in the field of advanced III-V semiconductors, and of the impact of that technology on future space communication systems. Following the program format outlined by NASA, the study was pursued in terms of distinct tasks, as follows:

Task 1: Space Communications Applications

Task 2: Present Status of Technology

Task 3: Competing Technologies

Task 4: Fundamental Limits

Task 5: Problems in Implementation

Task 6: Recommendations for Implementation

In general, the investigations followed the plan developed in the original proposal. The two deviations from that plan concerned Tasks 1 and 2:

In Task 1, it was originally envisaged that future space communications systems could be identified on the basis of existing planning documents in sufficient detail to make possible the development of broad component "specifications," which would then serve as the background for Tasks 3, 4, and 5. This did not prove possible, since few plans offering sufficiently precise projections - in terms of frequency, concept, or system architecture - could be gleaned from the literature. Thus, instead of very specific component specifications, the needs identified in Task 1 were broader in scope, based on subjective projections of what technical innovations future systems might contain, either in terms of evolutionary progress from present thinking (e.g., "large structures") or as a result of basic changes in technical approach (e.g., "reconfigurable transponder").

Although originally planned as one of a number of "coequal" tasks, it became apparent that Task 2 (Present Status of Technology) could well serve as the single most important task, central to the study. We therefore allowed (with the concurrence of the contract monitor) the expansion of this effort in terms of both scope and allocated time. Section III is the comprehensive report of this expanded investigation.

Portions of this report (Sections II, III, IV, and V) have been presented in summary form at program review meetings at NASA/LRC and RCA Laboratories.

The various tasks were treated as separate studies by different investigators, as planned in the proposal. The findings for each task are presented in the following sections of this final report.

II

Task 1: Space Communications Applications

Attempts to predict technology trends are always difficult, particularly when the technology area is not mature, so that inventions (which obviously cannot be used in the projections) may dramatically change the course of its application and the directions of the technology itself. In the microwave area, this has been the case in the past, and the solid-state microwave field has been particularly elusive in terms of accurate predictions of trends in component development and applications.

It was the original intent to define, through the effort of Task 1, future space communications applications in sufficient detail to enable the development of component specifications - in terms of such parameters as frequency, function, mechanical characteristics and reliability - to serve as a basis for the Study. This did not prove feasible because of the lack of existing planning information on which such specification could be based. It was, of course, possible to invent "future systems" solely for the purpose of defining the components they might require. Based on the program objectives, as discussed during an early planning meeting with NASA/LRC personnel (in September 1981), namely, the preparation of recommendations for NASA sponsorship of new broad technology development programs (rather than those for specific components) for future space communication systems, it was judged more productive to attempt to project future systems in terms of their more general characteristics and to extract new technological needs for systems so defined.

As a result of literature searches, meetings with personnel from technical and operating activities both within and outside RCA, and interviews with various individuals, the projections of future systems fell into three main categories:

(1) Systems that were logical extensions of systems currently being planned for the 1985-1995 time slot.

(2) Systems in which new types of modulation schemes would be employed.

(3) Novel system concepts that may strongly impact future space communications.

A. EVOLUTIONARY SYSTEMS CONCEPTS

Systems that are likely to evolve from those currently being planned for deployment in the 1985-1995 time frame are summarized in table II-1.

1. Large-Antenna Systems

Large structures for space are prominent in the present planning of both U.S. and European communications specialists. Some of these structures are merely assemblages of "normal" antenna and transponder systems. However, those containing large, multielement antennas may require extensions of known technologies. As indicated in Table II-1, large, multielement antennas may be used to generate fixed, switched, or steerable microwave beams; to reduce unwanted sidelobes; or simply to combine power from a multitude of low-power transmitting elements. From the component standpoint, such antennas will require sophisticated solid-state modules with self-contained control elements (phase, amplitude) as well as sophisticated on-board switching arrangements, most probably with dual-gate III-V-compound FETs.

2. Satellite Crosslinks

Crosslinks between communications satellites in geostationary orbit have been proposed, both for simplifying transmission problems (elimination of multihops between Earth stations and spaceborne transponders) and achieving better security. Millimeter-wave or laser systems (see note below) are attractive, since atmospheric attenuation problems that may preclude the use of such a system in Earth links are nonexistent in space, relatively low power levels may suffice, and well-behaved beams can be generated with antennas of "reasonable size."

In the millimeter-wave range, reliable transmitter and receiver components will be required, and the hybrid-vs-monolithic tradeoffs will have to be evaluated once the specific frequencies are selected. For laser-based systems, component development extending present research into solid-state lasers, radiation-resistant lasers, modulation volumes, and - most important - power-combining techniques would have to be pursued.

Note: Although laser systems are beyond the scope of this study, they are briefly mentioned here because (a) they functionally overlap the study's areas

TABLE II-1. EXTENSIONS OF CURRENTLY PLANNED SYSTEMS

Feature	Function	Component Requirement
1. Large Antenna	• In-space power combining • Beam steering • Low-sidelobe beams • Switched beams	Solid-state antenna module: Conformal Include gain/ power control Include phase shifter Linearity defined Microwave switch: Multirow/port Remotely activated Dual-gate FET
2. Crosslink Systems	• Improving frequency reuse	Millimeter-wave module: Hybrid-monolithic Wideband Fixed power/gain Laser transmitter/ receiver: Solid-state laser Modulator/ detector Receiver components Laser power-combining techniques
3. SSB Voice Systems	• Spectrum utilization	Superlow-noise mixers
4. Aerostat Systems	• Local coverage by captive platform	Conformal antenna modules: Power/cost tradeoff Near-field (tubular) beam generation
5. Digital Systems	• High-speed signal processing	Gigabit logic components

of technical concern and (b) the solid-state lasers do use III-V materials.

3. Single-Sideband (SSB) Systems

SSB wire systems are attractive from the standpoint of spectrum utilization. As indicated, special mixers may be required for such systems.

4. Aerostat Systems

It has been proposed that captive platforms hovering over a densely populated area (e.g., a large city) would be used to advantage for local traffic, possibly in an interactive mode of operation. For this type of system, a conformal antenna array would likely be required, which would generate a near-field tubular beam with practically no "spill-over" beyond the coverage area.

5. Digital Signal Processing

Digital signal processing is likely to dominate space communication systems of the future. Logic components capable of operation at gigabit/s rates will be required.

B. SYSTEMS USING ADVANCED MODULATION CONCEPTS

Systems in which advanced modulation schemes will be utilized are summarized in table II-2. Spread-spectrum techniques are particularly important in light of the recent FCC Notice of Inquiry (Gen. Docket No. 81-413, Sept. 15, 1981). Depending on the type of system used in a particular communications satellite (multicode vs single-code; frequency hopping between transponders; the type of multiplexing arrangements employed), special filters, switches, and subsystems may well be needed. In particular, the amplifier linearity requirements of a single-carrier spread-spectrum channel need to be studied and defined, and a transponder containing such amplifiers needs to be configured and tested.

Quadrature-amplitude (Q-A) modulation may provide a means of significantly increasing the use of an allocated frequency slot. This type of modulation scheme will depend on the technical feasibility (and economics) of developing "superlinear" power amplifiers without incurring unacceptable penalties in power-added efficiency. This problem - the tradeoff between bandwidth and linearity on the one hand and efficiency on the other - will require thorough theoretical and experimental exploration.

TABLE II-2. SYSTEMS USING ADVANCED MODULATION CONCEPTS

Feature	Function	Component Requirement
Spread-Spectrum Multiple-Carrier Techniques	• Better spectrum utilization A/J (military)	Special amplifiers (broadband) of still undetermined linearity
		Special transponder components (filters, multiplexers)
		Multicode single-code devices
		Switching between transponders for frequency hopping
Q-A (Quadrature-Amplitude) Modulation	• Better frequency utilization	"Superlinear" amplifier development
		Gigabit-rate signal processing

C. NEW CONCEPTS IN COMMUNICATION SATELLITES

The concept of a "reconfigurable transponder" (see table II-3) may prove to be of considerable significance in future systems. The idea involves adapting a spaceborne transponder for the transmission of particular signals upon command from a control station on Earth. Thus a transponder may be reconfigured to carry a TV signal rather than a multiplicity of voice signals; it may be changed from a single-carrier unit to a multicarrier transmitter; or the power it transmits may be adjusted upward or downward for a particular transmission. The basic technology for such an arrangement may already exist: Dual-gate FETs are capable of amplitude/gain control over wide ranges, while bandwidth control may be achieved by the use of switched filter stages or electronically adjustable active filters.

The other new concept that may be applied in communication satellites of the future is that of the RCA Solar Microwave Array Technology (SMART), also called "solar blanket" or "sandwich concept" in the past. The idea involves a structure having solar cells on one side and microwave modules on the other, so that dc power distribution problems are eliminated, graceful degradation of the antenna array is achieved, and different beam shapes can be obtained by proper phasing of the antenna element clusters. The major problem with such an arrangement is the necessity of simultaneously maintaining solar illumination on the solar-cell side of the structure and orientation toward the Earth (or another satellite) of the microwave-module side. Mechanical solutions to this problem have been worked out, but they will need detailed design and experimental verification. As to the SMART concept itself, extensive technology development for a specific application will be required. An intersatellite link operating at a millimeter wavelength and requiring relatively little power may well serve as the test structure for demonstrating the feasibility of this approach.

TABLE II-3. SYSTEMS USING NEW CONCEPTS

Feature	Function	Component Requirement
"Reconfigurable Transponder"	• Multifunction satellites	Electronic control of transponder bandwidth
		Electronic control of transponder power
		Electronic control of transponder linearity (gain)
		Constant-efficiency devices
SMART (Solar Microwave Array Technology)	• Simplify thermal, voltage-distribution, and power-combining problems of large arrays	Back-to-back solar-cell/ microwave module
		Reconciliation of solar/ microwave orientations

D. BIBLIOGRAPHY

AF Space Division: Technology Roadmap Industry Review, May 15, 1981.

American Institute of Aeronautics and Astronautics: Projected Space Technologies, Missions and Capabilities in the 2000-2020 Time Period, Jan. 14, 1981.

Bekey, I., Mayer, H. L., and Wolfe, M. G.: Advanced Space System Concepts and Their Orbital Support Needs (1980-2000), Vol. I: Executive Summary. The Aerospace Corporation, Dec. 1976.

Calio, A. J., et al.: NASA Space Communications Program Hearings - U.S. House of Representatives Subcommittee on Space Science and Applications, July 8-9, 1981.

Clark, J. F.: Planning for the Broadcasting-Satellite Service in the USA. IAF-81-70.

Corey, B. J.: 30/20 GHz Satellite Switching Matrix Development. NASA Industry Briefing, May 7, 1981.

Deane, C., and Moore, J.: Study to Determine the User Requirements for a Mobile Satellite System Operating in the 806-890 MHz Band. Canadian Department of Communications, Woods Gordon, Mgmt., Consultant and KVA, Contractor, Report No. DOC-CR-SP-81-025.

Drysdale, J. K.: Video 90. RCA Engineer, vol. 26, no. 7, July/Aug. 1981, pp. 12-19.

Fabis, B. F.: Some Aspects of the German TV-SAT System. Acta Astronautica, vol. 8, no. 7, 1981, pp. 775-785.

FCC Notice of Inquiry, Gen. Docket No. 81-413: In the Matter of Spread Spectrum and Other Wideband Emissions.... Sept. 15, 1981.

Ford Aerospace & Communications Corporation: Spacecraft IF Switch Matrix for Wideband Service Applications in 30/20 GHz Communication Satellite Systems. Final Report, Contract No. NAS3-22501, May 1, 1981.

Future Systems, Inc.: Pocket Data Communications. 1981.

Gabriszeski, T., et al., Western Union Telegraph Co.: 18/30 GHz Fixed Communications System Service Demand Assessment. NASA CR 159546, Vols. I, II, and III, July 1979.

Gamble, R. B., U.S. Telephone and Telegraph Corp., ITT: 30/20 GHz Fixed Communications Systems Service Demand Assessment. NASA CR 159619, Vol. I: Executive Summary and NASA CR 159620, Vol. II: Main Report, Aug. 1979.

Gamble, R. B., et al., UST and TC, ITT: Market Capture by 30/20 GHz Satellite Systems. NASA CR 165231, Vol. I: Executive Summary and NASA CR 165323, Vol. II: Final Report, April 1981.

Herdan, B. L.: The European Large Telecommunications Satellite Programme: Demonstration Mission and Future Perspectives. IAF 81-68, Sept. 1981.

Hughes Aircraft Co.: V-Band Communications Amplifier. USAF Avionics Laboratory TRACE AF-WAL-TR-81-1035, April 1981.

Intelsat Annual Report, 1981.

Keigler, J. E.: Development of Satellite Systems for Domestic Communications; in KFAS: Innovations in Telecommunications. Kuwait, April 1981.

Koelle, D. E.: Economy of Small Reusable LEO Satellite Platform. IAF 81-225, Sept. 1981.

MacDonald, R. I., and Hara, E. H.: Optoelectronic Switching Matrices Break GHz Barriers. Microwave System News, Nov. 1981, pp. 97-104.

Morgan, W. L.: The Economics of Large Orbital Communications Systems. IAF 81-226, Sept. 1981.

NASA: Advanced Communications Satellite. Industry Briefing, May 6-7, 1981, Washington, D.C.

NASA: Space Systems Technology Model. Third Issue, Sept. 1981. (Executive Summary, Vols. IA, IB, II, and III.)

NASA: Space Research and Technology Program and Specific Objectives. Fiscal Year 1982.

Nippon Telegraph and Telephone Public Corporation: Research and Development of Track III Products. Commerce Publishers Daily, Dec. 8, 1981, p. 1.

Ponzet, A.: The French Broadcasting Satellite Program. IAF 81-74, Sept. 1981.

Ranvoisy, P., and Congouet, C.: Twin Satellites for Broadcast, Distribution and Communication Satellites Systems. IAF, Sept. 1981.

Rao, V. R., and Vasagan, R. M.: Apple-Indian Experimental Geostationary Communication Satellite. IAF '81, Sept. 1981.

Rector, W. F., III, and Bowman, R. M.: Global Satellite Communications System Using Geostationary Platforms. IAF 81-52, Sept. 1981.

Rogers, J., and Reiner, P.: 30/20 GHz Net Accessible Market Assessment. Western Union Telegraph Co., NASA CR 159837 and Appendix, Feb. 1980.

Rutgers, The State University of New Jersey: Telecommunications in the Year 2000. Nov. 17-20, 1981.

Sivo, J., et al.: NASA Lewis Research Center Industrial Briefing for 30/20 GHz Communications Project, Nov. 1980.

Stofan, A. J., et al.: NASA's Next Decade in Space. National Space Club, June 1981.

Texas Instruments: 20 GHz POC FET Power Amplifier Design. NASA Industry Briefing, May 1981.

Tong, D.: Study of Potential Impact of Large Multifunctional Space Platforms on Canadian Satellite Systems. Canadian Department of Communications Contractor (Canadian Astronautics Limited), Report No. DOC-CR-SP-80-008, July 1980.

TRW: 30/20 GHz Spacecraft GaAs FET Solid State Transmitter. Task II Completion Report, Contract No. NAS3-22503, April 1981.

United Nations Outer Space Affairs Division: Review and Projection of Space Technology, Background Paper. Jan. 31, 1981.

III

Task 2: Present Status of Technology

This section reviews the present status of III-V compound technology as it pertains to space communications systems. Emphasis has been placed on materials and discrete device technology since this forms the basis of microwave and digital integrated circuits. We have attempted to project future trends and have included a section on the exploratory development that is being undertaken in many laboratories.

The field of III-V compound microwave and gigabit-rate logic technology is currently very active. We have tried to include most of the results published up to early 1982. The trends discussed in this section are believed to be reasonably accurate; some better data may be available that we have missed. The extensive literature references provided are representative rather than complete or in historical order.

We believe that a major change is occurring in III-V compound technology. The major attraction of GaAs for microwave and gigabit-rate digital ICs until recently was mainly the superior semiconducting properties of GaAs and not its fabricating technology. New materials growth techniques such as molecular-beam epitaxy (MBE) and metal/organic chemical-vapor deposition (MOCVD) permit an unprecedented complexity, flexibility, and diversity in III-V compound growth. This technological strength, in addition to superior fundamental semiconducting properties, promises to make III-V compound research an exciting and useful endeavor in the mid-1980 to 2000 time frame.

A. MATERIALS TECHNOLOGY

1. Bulk Growth of Semi-Insulating (SI) GaAs and InP

The demand for GaAs substrates is currently being dictated by microwave and gigabit-rate logic applications. With the development of the GaAs FET and the evolution of technology toward an integrated-circuit format for both linear microwave and gigabit-rate logic applications, this demand is for the semi-insulating (SI) form of GaAs. These developments have led to a renewed interest in both the technology of bulk SI GaAs growth and the physics of SI GaAs. The first conference on semi-insulating III-V materials, attesting to the research activity in this area, was held in Nottingham, England, in 1980 [1]; the second conference was held in April 1982, in France.

Before these developments, the driving force behind the bulk growth of GaAs was the demand for highly conducting n^+ substrates for light-emitting diodes (LEDs). The LED substrate requirements are much less stringent than for microwave applications, and the most popular methods for substrate growth were the "boat growth," horizontal Bridgman, or gradient-freeze techniques. Some liquid-encapsulated Czochralski (LEC) substrates were also available but the boat-growth methods were more economical and adequate for LEDs and microwave devices that use n^+ substrates (discrete transferred-electron devices and IMPATTs). The inadequacies of n^+ substrates could also be overcome very easily by the epitaxial growth of an n^+ "buffer" layer. At that time the demand for SI GaAs was not very high; the material was used for making samples for Van der Pauw measurements and for some research carried out on GaAs FETs.

With the development of the GaAs FET and the demonstration of its potential for low-noise and medium-power amplifiers, linear ICs, and gigabit-rate logic ICs, the demand for SI GaAs increased. Until recently, most SI GaAs was obtained by adding sufficient Cr to the melt to render the material insulating. Boat-grown, Cr-doped GaAs was not very reproducible, being sometimes semi-insulating, sometimes p type, and often with different spatially segregated regions. Other problems include surface conversion during epitaxial growth, fast-diffusing deep-level acceptors, etc., leading to poor electron mobility particularly at the active layer-substrate interface. The development of ion implantation into SI GaAs has put even more stringent requirements on the uniformity and stability of the substrate. This has led to increased effort on both Bridgman and LEC systems to improve substrate quality. A development that promises to have a major impact on SI GaAs growth is the Melbourn high-pressure LEC puller manufactured by Metals Research, Ltd., England. This crystal puller was developed at Metals Research and relies strongly on the research effort at the Royal Signals and Radar Establishment, England.

a. Substrate Requirements

Before a brief description of current research in bulk GaAs growth, the major requirements on SI GaAs substrates for microwave and gigabit-rate logic are presented.

(1) The most obvious requirement is that an ample supply of reproducible, large-area (~80-mm diam) substrates be available. A number of systems houses (captive supply) and vendors have invested in LEC equipment, and a larger

quantity of SI GaAs substrates is now becoming available. Despite promising indications about improving quality, problems still occur intermittently.

(2) The resistivity of the SI GaAs substrates must be sufficiently high ($\rho \geq 10^7\ \Omega \cdot cm$) and must not degrade after processing steps such as epitaxy and/or ion implantation and annealing. Current research indicates that this requirement implies that the background shallow donor and acceptor levels should be sufficiently low ($\leq 10^{15}\ cm^{-3}$) [2,3].

(3) Obviously the crystal perfection should be high, especially insofar as growth imperfections such as stacking faults, inclusions, precipitates, twins, low angle boundaries, etc. are concerned. Intuitively, one would also like the dislocation density to be as low as possible. The effect of dislocation density on device yield and performance has not yet been unambiguously established.* It is plausible, however, that dislocations may act as sinks for impurities and contribute to anomalous rapid diffusion processes and impact device and IC reliability [4]. Currently, boat-grown and LEC SI GaAs have dislocation densities in the ranges of 10^3-10^4 and 10^4-$10^5\ cm^{-2}$, respectively.

(4) The substrate should allow the generation of abrupt carrier profiles with high and controlled electrical activation and high carrier mobility by ion implantation and anneal. This operational requirement is dependent upon the previous factors and is also a function of the implantation and annealing technique.

(5) Another operational requirement is that there be minimum trap-related phenomena at the active layer-substrate neighboring devices [7]. Since deep traps are used to render the substrate semi-insulating, some backgating may be unavoidable and may have to be compensated for by device and/or circuit design. This is hampered by the fact that these effects vary from ingot to ingot. In fact, if the crystal is grown in a direction different from <100> and then (100)-oriented wafers are cut, it will consist of regions freezing at different times during growth. Trap-related effects may then vary across a substrate wafer. This requirement is thus related to the previous ones, particularly that of reproducibility.

---

*A recent publication, Nanishi et al., Jpn. J. Appl. Phys., vol. 21, no. 6, June 1982, pp. L335-L337, indicates that nonuniform threshold voltage distribution in GaAs FETs is related to the dislocation density distribution.

(6) As the quality of SI GaAs substrates improves and more complex ICs emerge, new requirements arise. One requirement is the availability of large-area (60- to 80-mm diam) regularly shaped wafers. Most semiconductor processing equipment is developed for Si-based devices and handles circular wafers. Boat-grown SI GaAs usually has a characteristic D shape; some rectangular crystals are now becoming available [8]. <111>-pulled LEC SI GaAs when cut to (100) has an elliptical shape. Recent work has shown that LEC <100>-pulled crystals of high quality can be grown to yield circular (100) wafers 60-80 mm in diameter, that can be provided with a <110> flat - looking much like a conventional Si substrate [2,3].

The above fundamental and operational requirements must be met before a viable and cost-effective microwave and gigabit-rate technology can be realized. The theoretical understanding of SI GaAs must improve so that growth technology can be refined. As discussed by White [9], this is a circular requirement, since reproducible SI GaAs of good crystal quality must be available before one can characterize the wafers with confidence. Empirical research has now reached a stage wherein a more systematic approach to the development of SI GaAs growth can be undertaken.

b. Bulk Growth of SI GaAs

The bulk growth of SI GaAs is described briefly to provide a baseline for discussing current research. This description is not meant to be exhaustive.

The intrinsic carrier concentration in GaAs at room temperature (300 K) is about $2\text{-}3 \times 10^{7}$ cm^{-3} [10]. Barring some unforeseen breakthrough, the control on impurities required to grow intrinsic material does not appear feasible in the near future. In fact, the control on shallow donors and acceptors at a 10^{15}-cm^{-3} level to achieve so perfect a compensation so as to pin the Fermi level at midband is unlikely. The most common method for realizing SI GaAs would still be the use of deep acceptors (donors) to more than account for the net residual shallow donors (acceptors). The most common deep-level impurities are Cr acceptors, O donors, and deep-level native defects (e.g., EL2 [2]). The theoretical modeling of SI GaAs is complex because of the multivalent nature of impurities such as Cr [9,11,12]. The invited papers at the 1980 Conference on Semi-insulating III-V Compounds deal with these problems in some detail with many references [1]. At any rate, it is clear that one major requirement is to minimize the net shallow impurity levels in bulk-grown GaAs to 10^{15} cm^{-3} or lower so that a minimum, controlled amount of deep-level impurities can be introduced to obtain stable and reproducible SI ingots. The most common shallow-level impurity is Si, usually from the quartz growth equipment.

GaAs growth technology is considerably more complex than that of Si since one is dealing with binary-phase equilibria and a highly volatile (As) component. Accurate control of As vapor pressure is necessary to maintain GaAs stoichiometry and crystal perfection. The growth process consists of two basic steps: First, GaAs is compounded by the use of ultrahigh-purity Ga and As. This is followed by single-crystal growth.

i. Boat Growth: In this method the compounding is carried out by reacting 6/9's Ga with As vapor (generated from 6/9's As) at elevated temperatures in sealed high-purity quartz ampoules. Typically, an As reservoir at one end of the ampoule is heated to about 600°C, resulting in about 1 atm of As vapor pressure. The control of As vapor pressure is essential for obtaining stoichiometric GaAs [13]. The As vapor reacts at about 1260°C with Ga metal located in a quartz boat at the other end of the ampoule. After complete reaction of Ga, single-crystal growth is started by programmed cooling (gradient-freeze technique) or by moving either the furnace or ampoule to provide the requisite temperature gradient (horizontal Bridgman). The melt can also be seeded appropriately. Figure III-1 is a schematic of a basic horizontal Bridgman system. Many variants of the basic system, such as the use of three temperature zones, are employed. The Bridgman technique can be automated easily, requires minimum supervision, and requires less capital outlay than the LEC system.

The major problem in boat-grown GaAs is due to the contact between the melt and the boat. Quartz is the most commonly used material leading to Si contamination. Si contamination is also a function of As pressure, decreasing with increasing partial pressure [14]. The dislocation density, however, appears to increase with As partial pressure [15]. Doping of the boat-grown material is carried out by adding the dopant (Cr for SI GaAs) to the Ga melt. In a later subsection we will compare the residual impurities in boat-grown and LEC SI GaAs. As mentioned earlier, boat-grown GaAs is irregular in shape (i.e., D-shaped), but some rectangular <110> ingots are also being grown [8]. Table III-1 lists some vendors of boat-grown SI GaAs and the characteristics of their material taken from specification sheets.

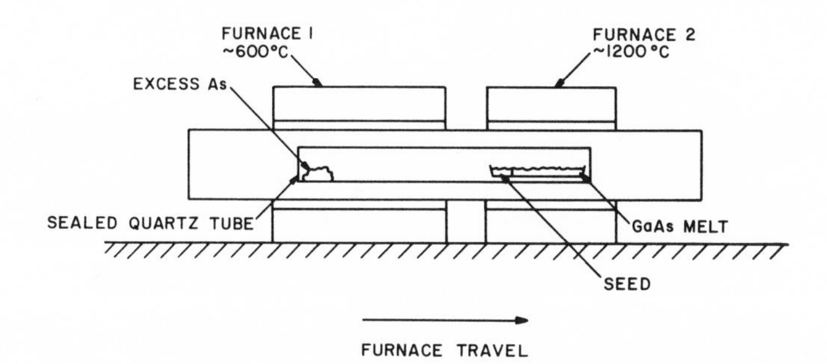

Figure III-1. The horizontal Bridgman technique for GaAs with the temperature profile in the double furnace.

ii. Liquid-Encapsulated Czochralski (LEC) Growth: In the Czochralski method, a crucible is used to contain the melt. A seed held by a rod is dipped partially into the melt and, after equilibration, the seed is pulled by the rod (hence the term "pulled crystal") while the rod is rotated at an appropriate rate. Sometimes the crucible is rotated too. Rotation of the seed and crucible is employed to optimize compositional and thermal homogenization of the melt and to control crystal geometry. This basic configuration must be modified to grow crystals that have a volatile component (e.g., III-V's) by completely covering the melt with an inert-liquid encapsulant and maintaining an inert-gas pressure higher than the equilibrium dissociation pressure of the most volatile melt constituent. The liquid-encapsulated Czochralski (LEC) technique was first applied to the growth of PbTe and PbSe by Metz et al. [16] in 1962 and employed by Mullin et al. [17] in 1965 for III-V compounds. The requirements of the encapsulant are that it should be less dense than the melt (e.g., it should float on the melt) and be optically transparent and chemically stable. Boric oxide, B_2O_3, has proved to be the most successful material for III-V compound growth.

The conventional LEC technique for the growth of GaAs consists of two separate steps: viz., compounding and crystal growth [13]. The increased handling involved in this technique leads to considerably higher impurities than those found in boat-grown material. The development of high-pressure LEC

Table III-1. COMMERCIALLY AVAILABLE BOAT-GROWN SI GaAs*

Vendor	Resistivity (Ω·cm)	Dopant	Shape/Size	Dislocation Density (cm^{-2})	Residual Impurities (ppma) Si	B	Cr	Fe	Vendor Comments
Crystal Specialties, Inc.	10^6-10^8	Cr	D/54 mm x 35-50 mm; Rectangular/ 50-55 mm x 35-45 mm	.... Not supplied					Horizontal Bridgman - quartz apparatus; <111> grown
Sumitomo Electric Industries, Ltd.	10^7-10^8	Cr-O	D/54 mm x 27 mm	5 x 10^3-10^5 Various grades	0.3-1	Not determined	-	0.01-0.03	Three temperature zones. Horizontal Bridgman - quartz boat; type IV material - no thermal conversion after 800°C anneal - circular wafers under development - <111> grown

*Data supplied by vendors in response to questionnaire. Some vendors did not respond.

pullers (The Melbourn, developed by Metals Research, Ltd., England) has led to a direct synthesis process wherein compounding is also carried out under a B_2O_3 melt under an inert-gas pressure of 55-60 atm. In situ compound synthesis is possible from elemental Ga and As since B_2O_3 melts before significant As loss occurs (∿450°C). Compound synthesis occurs exothermally at 820°C under 55-60 atm of inert-gas pressure to prevent dissociation. Crystal growth can then be started from the stoichiometric melt by seeding and pulling the crystal.

The Melbourn LEC equipment is being adopted by many substrate vendors and systems houses for the growth of GaAs, GaP, and InP substrates. The puller (fig. III-2) consists of a resistance-heated, 6-in.-diam crucible capable of holding up to 10 kg of charge, and can be operated at pressures up to 150 atm. A closed-circuit TV system is used to view the melt within the high-pressure vessel. A high-sensitivity weight cell continuously weighs the crystal during growth and provides a differential-weight signal that can be used for manual diameter control. An automatic diameter-control technique using a diameter-control die made of Si_3N_4 (called the "coracle") floating on the melt has been developed for <111>-oriented growth. This process cannot yet be used for <100>-oriented growth because of twinning problems. <100>-grown crystals are centerless ground, cut into round (100) substrates, and provided with a <110> flat [2,3]. Circular substrates with diameters ranging from 50 to 80 mm have been grown [2,3] and are available commercially.

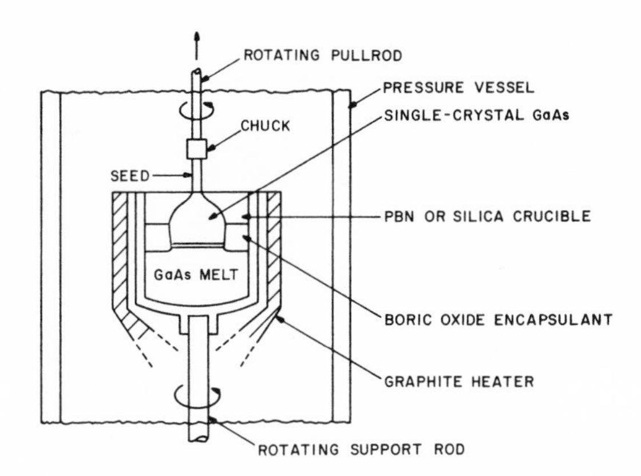

Figure III-2. Schematic of LEC system.

The use of quartz crucibles for compounding GaAs and growing single crystals leads to substantial incorporation of Si. Si can act as either a shallow donor or acceptor (i.e., be incorporated on a Ga or As site) in GaAs, depending on the specific growth conditions, and can prevent the attainment of high-resistivity undoped and/or Cr-doped material. A recent development is the use of either pyrolytic boron nitride (PBN) crucible liners or PBN crucibles. Some researchers claim that the residual impurities in PBN-grown LEC SI GaAs are lower than in quartz systems [3], while others indicate that there is no significant difference [2]. In principle, a PBN crucible is more desirable since it removes a source of Si.

It has been reported that large-diameter (~80 mm) LEC GaAs crystals are usually characterized by radially nonuniform dislocation distribution with a maximum in the 10^4- to 10^5-cm^{-2} range [2,3]. <111> and <100> ingots free from microstructural defects such as twins, inclusions, and precipitates have been achieved. Coracle-controlled <111> crystals with ±4% diameter variation have been realized [2,3]. The major dislocation generation mechanism is believed to be due to thermal stresses [18], and most dislocations are in the region close to the outer periphery of the boule [18]. Dislocation-free small-diameter (1.5 cm) crystals have been reported [19]. A current research goal is to reduce the dislocation density.

Papers from laboratories where both LEC growth and device fabrication research are being carried out claim that the undoped and Cr-doped LEC substrates are superior to most boat-grown SI GaAs substrates insofar as direct ion-implantation yield is concerned [2,3]. Research at laboratories that use commercial direct-synthesis LEC material, however, does not provide clear-cut answers. This is perhaps indicative of the fact that the quality of commercially available LEC substrates is not yet as high as that in research laboratories. Table III-2 lists some vendors that sell LEC SI GaAs. The large investment in the LEC process is expected to pay dividends in about the next five years.

c. Residual Impurities

It is desirable that the SI GaAs substrates have low shallow impurity level density ($\leq 10^{15}$ cm^{-3}) so that a minimum controlled density of deep levels is required to render the material insulating. Since there is considerable movement of deep level causing impurities (e.g., Cr) during processing (e.g.,

Table III-2. COMMERCIALLY AVAILABLE LEC SI GaAs*

Vendor	Resistivity (Ω·cm)	Dopant	Shape/Size	Dislocation Density (cm^{-2})	Residual Impurities (ppma) Si	B	Cr	Fe	Vendor Comments
Cominco American	$>10^7$	None	Circular/ 50-, 60-, 75-, and 60-, 75-, and 100-mm diam with flats	$0.5\text{-}5 \times 10^5$	......Not supplied......				In situ compounding, PBN crucible, <100> axis, Melbourn puller
Microwave Associates	$>10^7$	None	Circular/ 75-mm diam	$\leq 10^5$	......Not supplied......				
	$>10^7$	Cr	Circular, 75-mm diam	$\sim 10^5$					
Metals Research	$10^7\text{-}10^8$	None	Circular/ 75- to 100-mm diam with (110) flat	$\sim 10^5$	0.01	0.5	Not determined	Not determined	LEC, both quartz and PBN, in situ compounding
	$10^7\text{-}10^8$	Cr	Circular/ 75- to 100-mm diam with (100) flat	$\sim 10^5$	......Not supplied......				Pulled <100> or <111> - customer option
Mitsubishi Metal Corp.	10^7	Cr	Circular/ 50- to 75-mm diam	$\sim 4 \times 10^4$	0.05	0.5	--	0.015	LEC quartz, developing PBN, <100> growth axis; no conversion, 850°C, 15-min anneal
Wacker Siltronic Corp.	$10^6\text{-}10^8$	Cr,O	Circular/ 40-mm diam with (110) flat	$0.5\text{-}1 \times 10^5$	......Not supplied......				LEC quartz and PBN, both separate and in situ compounding
Sumitomo Electric Ltd.	$10^6\text{-}10^7$	Cr-O	Circular/ 40- to 55-mm diam	$0.3\text{-}1 \times 10^5$	0.1	0.03	0.1	0.01	LEC quartz, Sumitomo puller, PBN under development, separate compounding, in situ under development, <100> growth axis

*Data supplied by vendors in response to questionnaire. Some vendors did not respond.

due to implantation and annealing), the presence of a large amount of shallow impurities can cause material type conversion and its attendant problems [20]. There is, therefore, a large effort in many laboratories to measure the atomic concentration of residual impurities by secondary ion mass spectrometry (SIMS), and the electrical concentration of deep-level impurities by the many forms of deep-level trap spectroscopy (DLTS) [2,3].

Table III-3 summarizes SIMS measurements made on SI GaAs and reported in the literature. The results for LEC/PBN, LEC/quartz, and boat-grown material are included. These represent data reported early in 1981 and are thus indicative of the quality of the best material available in 1980. We believe, however, that this is not indicative of typical commercial material quality. Note that data in table III-3 include results from captive material growth centers and (unidentified) commercial sources. The preliminary conclusions to be drawn from table III-3 are as follows:

(1) The Si content in direct-synthesis LEC SI GaAs is fairly low for both PBN and quartz-crucible-grown material. In general, the material grown in PBN crucibles has lower Si than quartz-crucible-grown GaAs. The Si content of commercially available LEC SI GaAs appears to be higher than that of material grown in research laboratories by a factor of 5 or more. It must be noted, however, that the sample size is relatively small.

(2) The Si content in boat-grown GaAs is higher than that in LEC GaAs grown at research laboratories. The situation for commercially available LEC and boat-grown material is not so clear cut.

(3) S, a group VI atom (donor), is a significant impurity in all bulk-grown GaAs. S is a common trace impurity in high-purity As.

(4) A high concentration of B is observed in LEC/PBN material. This B is believed to be electrically neutral and does not contribute significantly to ionized impurity scattering [21]. It is interesting to note that one laboratory [2] quotes B concentration for LEC/PBN material almost two orders of magnitude lower than another [3]. Boat-grown material has a considerably lower B concentration.

(5) The residual Cr concentration in undoped LEC/PBN is close to the SIMS detection limit. One example of a commercial LEC/quartz undoped GaAs has about 10^{15} cm^{-3} of Cr. LEC/quartz undoped material grown at research laboratories, again, has Cr levels close to the SIMS detection limit.

Table III-3. TYPICAL RESIDUAL IMPURITY LEVELS IN SI GaAs (1980-1981)

Crystal Type	Concentration: atoms x 10^{15} cm^{-3}							Dopant	Ref.	Comments
	Cr	Si	S	B	Mn	Fe	Mg			
LEC/PBN*	0.4	0.5	1	500	0.8	NM**	0.7	None	[3]	NCV†
	0.2	1.9	5	3.4	1	2.3	3.2	None	[2]	NCV
	5	0.4	3	200	0.8	NM	4	Cr	[3]	NCV
LEC/Quartz	0.8	1	2.5	0.7	1	2.5	4	None	[2]	NCV
	0.5	0.8	1	2	0.6	NM	0.3	None	[3]	NCV
	1	40	4	1	1	NM	2	None	[3]	FCV††
	20	1	2	1	0.8	NM	0.2	Cr	[3]	NCV
	10	0.8	1	2	0.7	NM	0.5	Cr	[3]	NCV
	20	5	6	2	3	NM	2	Cr	[3]	FCV
	60	8	4	2	2	NM	3	Cr	[3]	FCV
Boat-Grown	20	8	3	0.1	1	NM	2	Cr	[3]	FCV
	40	6	6	0.6	1	NM	1	Cr	[3]	FCV
	20	5	3	1	2	NM	1	Cr	[3]	FCV
	20	5	4	0.2	<0.5	3	0.27	Cr	[2]	FCV
SIMS Detection Limit	0.4	0.3	0.3	0.01	0.7	1	0.3			

*Liquid-encapsulated Czochralski/pyrolitic boron nitride.
**Not measured.
†Not commercial vendor (average values).
††From commercial vendor.

(6) The Cr content in Cr-doped boules found suitable for ion implantation generally lies in the range $2\text{-}4 \times 10^{16}$ cm^{-3} at the seed end and increases up to 10^{17} cm^{-3} toward the tang end. This points out the desirability of pulling <100> ingots; <100> Cr-doped wafers cut from <111> pulled ingots will have varying Cr concentration across their area in addition to variations from wafer to wafer.

(7) The lower residual donor content in LEC/PBN materials allows the use of lower Cr doping (3×10^{15} cm^{-3} seed, 6×10^{15} cm^{-3} tang) to obtain SI GaAs suitable for ion implantation [3].

(8) It is as yet not possible to obtain an accurate evaluation of the O content in SI GaAs by SIMS. Tentative values are in the mid-10^{16}-cm^{-3} range.

In summary, the residual impurities in direct-synthesis LEC material appear to be decreasing as time goes on. The commercially available material is not of as high a quality as that reported by research laboratories. The major research effort in bulk-growth technology is to decrease residual impurity levels, decrease dislocation density, improve the uniformity and yield of SI GaAs substrates, and develop operationally simple techniques for the assessment of substrate quality. In addition, there is renewed interest in understanding the physics of SI GaAs since the material quality is improving to the extent that meaningful results can be obtained. This will, in turn, help in improving SI GaAs quality.

d. Bulk Growth of InP

The bulk growth of single-crystal InP is much more difficult than that of GaAs. Yields have been low because of the tendency of LEC InP to twin (i.e., grow in two or more orientations in the same boule). The development of the crystal growth process for InP is considerably less advanced than that for GaAs and GaP, mainly because the latter two were initially financed by an expanding displays market. The technology of bulk InP growth has recently, however, improved to the point where reasonable yields can be obtained. The pioneering research was again carried out by Metals Research in UK. A review article by Rumsby et al. [22] presents the current state of the art.

While boat-grown single-crystal InP has been reported [23,24], most current commercially available InP substrates are grown by the LEC process. Separate compounding of InP charge is carried out [22,25]. Owing to the high decomposition partial pressure of P, synthesis of InP at the stoichiometric pressure requires a high degree of partial-pressure control for P to avoid explosions. Polycrystalline charge is usually compounded at a lower temperature in a manner analogous to horizontal Bridgman [25] or zone melting [22]. LEC growth is then carried out, usually in resistance-heated pullers of the Melbourn type. <100> crystals of 850- to 3000-g weight have been pulled [22].

Semi-insulating InP has been achieved by Cr-doping [26]. The solubility of Cr in InP is low, however, and very low background impurity levels must be achieved for successful Cr doping. Fe doping is a much less demanding alternative [27] and is almost universally used. Rumsby et al. [22] present typical residual-impurity data. Table III-4 lists the characteristics of Fe-doped InP that is commercially available. Circular (100) wafers up to 80 mm in diameter can be obtained. SI InP substrates are three-to-four times more expensive than SI GaAs substrates.

2. Epitaxial Growth

Epitaxial growth is perhaps the best-developed technology at this time for the realization of high-quality n and p layers of GaAs and other III-V compounds. Ion-implantation technology, because of its flexibility, is now challenging the supremacy of epitaxy, particularly in GaAs. This section will briefly describe the various epitaxial techniques currently in use for the growth of III-V compounds for microwave and gigabit-rate logic applications.

The epitaxial growth of III-V compounds dates from the early 1960s. Williams and Ruehrwein described the growth of GaAs and $GaAs_{1-x}P_x$ using the hydride vapor-phase-epitaxy (VPE) system in 1961 [28]. Liquid-phase epitaxy (LPE) of GaAs was reported by Nelson in 1963 [29]. In 1965, single-crystal growth of GaAs by the $AsCl_3$ process was reported by Knight and Effer [30]. In 1968, molecular-beam epitaxy (MBE) was first reported by Davey and Pankey [31], and the metal/organic chemical-vapor deposition of III-V's was reported in 1969 by Manasevit and Simpson [32]. Major advances in all of these epitaxial techniques have occurred since then. The driving force behind most developments in epitaxy was a need for device-quality material for microwave and electro-optic devices. The need to engineer suitable material for light sources and detectors covering a large range in wavelength (visible and IR) has led to the development of heteroepitaxial growth techniques for III-V compound ternary and quaternary alloy heterostructures lattice-matched to binary III-V substrates like GaAs and

Table III-4. CHARACTERISTICS OF COMMERCIAL SI InP SUBSTRATES*

Vendor	Resistivity (Ω·cm)	Dopant	Shape/Size	Dislocation Density (cm^{-2})	Residual Impurities (ppma)				Vendor Comments
					Si	B	Zn	Fe	
Metals Research	$\sim 10^7$	Fe	Circular/ 80-mm diam	$4\text{-}10 \times 10^4$	0.1	0.2	0.02	1-2	Separate compounding, <100> or <111> pulled
Sumitomo Electric Ltd.	$>10^6$	Fe	Circular/ 50-mm diam	$\sim 5 \times 10^4$	0.03	--	<0.03	0.8	Sumitomo LEC system, quartz crucible
Mitsubishi Metal Corp.	$\sim 10^7$	Fe	Circular/40- to 45-mm diam	$\sim 5 \times 10^4$	..... Not available				LEC quartz, separate compounding, <100> pulled
CrystaCom, Inc.	$\sim 10^7$	Fe	--	$\sim 10^4$	..... Not available				**<111> pulled crystal, 40-mm diam**
Varian Associates	$\sim 10^7$	Fe	--	5×10^4	..... Not available				LEC - no further details
	10^6	Cr	--	--			--		No further details

*Data supplied by vendors.

InP [33,34]. These alloy heterostructures may find potential microwave and gigabit-rate logic applications [35,36].

a. Vapor-Phase Epitaxy (VPE)

The VPE growth technique is currently the dominant epitaxial growth technique used in industry to obtain n- and p-type GaAs, InP, and ternary and quaternary III-V alloys. The VPE processes used can be subdivided into (i) chloride and (ii) metal/organic chemical vapor deposition (MOCVD) processes. The chloride processes are more developed than the MOCVD process but important gains are being made in MOCVD. VPE is a flexible process, and n and p layers varying in concentration from mid-10^{14} to 10^{19} cm^{-3} and thicknesses ranging from a fraction of a micrometer to 100 μm can be grown. VPE systems are relatively easy to scale up, and several production-oriented systems are available [37,38]. A large body of literature exists on the VPE of III-V compounds, and specialist conferences are held on the subject annually and biennially. A general tutorial review of VPE can be found in Volume III of the Handbook on Semiconductors [39].

VPE is, in many respects, still an empirical technology. There is a need for more basic knowledge of many different aspects of VPE, and a considerable amount of work is being carried out in universities and industrial laboratories throughout the world. Our discussion will be device-oriented and aimed at the specific microwave and gigabit-rate logic application.

i. Chloride VPE Systems: The two major chloride VPE systems of practical interest are as follows:

(1) the $Ga/AsCl_3/H_2$ (and $In/PCl_3/H_2$) system in which the $AsCl_3(PCl_3)$ vapor in H_2 carrier gas is passed over heated Ga(In) at about 850°C. Ga(In) is thus transported as GaCl(InCl) and reacts with As(P) liberated from $AsCl_3(PCl_3)$ to form GaAs(InP) on a substrate at about 700-750°C. This method produces layers of very low impurity concentration since only two high-purity chemicals are used. Very high purity Ga, In, $AsCl_3$, and PCl_3 are available. In some systems the Ga(In) source is replaced by high-purity single-crystal GaAs (InP).

(2) The hydride system in which the As(P) is supplied as $AsH_3(PH_3)$ and the group III metal is transported by reacting the metal with high-purity HCl. H_2 is used as the carrier gas. For growth of the binary III-V compounds, three high-purity source chemicals (e.g., Ga, AsH_3, and HCl for GaAs) are required,

resulting, in general, in higher amounts of residual impurities than are produced in the trichloride system. The gas-phase group III-to-group V ratio, however, can be controlled independently, leading to higher flexibility. This system is particularly suited to the growth of ternary and quaternary III-V alloys.

The chloride systems have not proved generally successful for the growth of Al compounds because of the reactivity of the $AlCl_3$ [40,41]. LPE and MOCVD are more suited for the growth of Al-containing compounds.

ii. Trichloride Process: Figure III-3 shows a schematic diagram of a "two-bubbler" trichloride system typically used for the growth of GaAs for microwave device applications. We will briefly describe the GaAs growth system because of its technological importance. The system for the growth of InP is very similar, and comments specific to InP will be interjected where necessary. The chemistry of this process has been described by Shaw [42], Hollan [43], DiLorenzo [44], and many others.

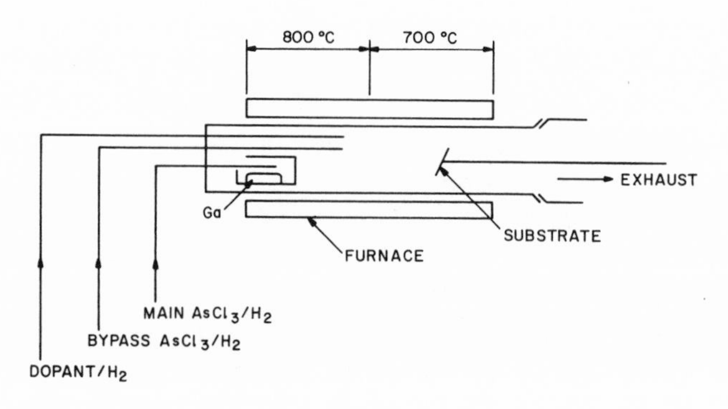

Figure III-3. Two-bubbler $AsCl_3/Ga/H_2$ CVD system [46].

The two-bubbler system first described by Nozaki et al. [45] and then Cox and DiLorenzo [46] is particularly suited for the growth of GaAs on SI GaAs. $AsCl_3$ is contained in two thermostatically controlled bubblers (typically at 18-20°C). High-purity Pd-diffused H_2 is passed through the bubblers to give a 10^{-3}-10^{-2} mole fraction of $AsCl_3$ in the gas stream. $AsCl_3/H_2$ from the "bypass" bubbler enters downstream of the Ga source. The reactor tube is fabricated from high-purity synthetic quartz, and the substrate is held at 700-750°C.

The major characteristics of this system are shown below.

(1) In the source zone, $AsCl_3$ decomposes to form As_2, As_4, and HCl. The HCl formed reacts with the source Ga (or GaAs) in a quartz boat to form GaCl. When a Ga source is used, As is initially taken up by Ga until it becomes saturated and a thin crust of GaAs forms upon it. This is known as source saturation and is a very important step. The Ga source must be completely covered by a GaAs skin, or the layer quality degrades drastically. The problems of source saturation were first described in detail by Shaw [42].

A similar source saturation step obtains for InP growth [43].

(2) The electrical purity of the layers grown is strongly dependent on the mole fraction of $AsCl_3$ in the reactor. This is the total $AsCl_3$ mole fraction, i.e., the sum of the main and bypass flows. At 10^{-3} mole fraction, the background doping (n type) is approximately 10^{16} cm^{-3}, while at 10^{-2} mole fraction the carrier level can be reduced to 10^{13} cm^{-3} in a well-designed system. This mole fraction effect has been shown to be due to the reaction of the HCl produced with the quartz (SiO_2) reactor tube [44,47]. The volatile chlorosilanes produced are reduced by H_2 near the substrates, leading to Si incorporation on donor (Ga) sites. Theoretical and experimental studies show that Si incorporation is proportional to p^{-n} $AsCl_3$ where n lies between 2 and 3 [47].

A similar mechanism obtains for the growth of InP, but rigorous steps must be taken to exclude O_2 and H_2O.

(3) The carrier concentration can be varied by mole fraction control or by the controlled addition of S, Si, Se, etc. A vast body of literature exists on dopant incorporation [47,48], surface morphology of layers [49], substrate orientation effects [50], etc.

(4) The two-bubbler method offers more flexibility than the conventional single-bubbler system [45,46]. When the bypass flow is much higher than the main flow, the substrate is etched in situ. At a lower bypass flow rate, growth occurs at a high $AsCl_3$ mole fraction and has low carrier density. When such a layer is grown on a Cr-doped SI GaAs substrate, the out-diffusion of Cr from the substrate renders this layer semi-insulating. By terminating the bypass flow and doping suitably, the active layer is grown [45,46]. Thus a

"controlled-compensation" method can be used to grow high-resistivity buffer layers for GaAs FET applications.

(5) High-resistivity buffer layers can also be grown by Cr-doping with CrO_2Cl_2 as the dopant source [51-53]. This method has some operational problems since Cr oxides deposit on the quartzware, causing devitrification. These problems can be solved by the use of appropriate liners of PBN or other inert material [52]. In an elegant method described by Cox and DiLorenzo, this deposit serves as the dopant source and Cr is transported by passing HCl over it [53]. A review of the epitaxial growth of SI GaAs can be found in reference 53.

(6) The technology for the growth of p-doped layers is not as well developed as that for n-doped layers. A major reason for this is that most GaAs microwave devices are majority-carrier devices, and the electron mobility in GaAs is higher than the hole mobility. The growth of highly doped p^+ layers for IMPATTs is relatively simple; however, the growth of moderately doped p layers (e.g., for double-drift IMPATTs) is not yet under control. Zn-doped p layers can be grown by use of the chloride, bromide, or iodide. These methods require precise temperature and gas-flow control, and heated lines to prevent condensation before the materials reach the reaction zone [54]. This makes growth of abrupt profiles difficult. Zinc alkyls can also be used to transport Zn. For low-doped p-layer growth, the generation of Zn iodide by passing HI over zinc arsenide in the reactor hot zone appears promising [54].

In summary, the trichloride process is fairly well understood, and single- and multiwafer reactors that can handle 60-mm-diam and larger substrates are in operation in many laboratories. Commercially made reactors are also available [37,38]. By proper reactor design, doping and thickness uniformities in GaAs layers of ±5% can be obtained; uniformities of ±1% in thickness and ±2.5% in doping have been reported [55]. High-quality InP layers can also be grown.

iii. <u>Hydride Process</u>: As indicated earlier, the hydride VPE synthesis technique that uses group V hydrides and group III chlorides by reacting the metal with HCl is a very versatile process. Among a large number of III-V binary, ternary, and quaternary alloys grown by this process are, for example, GaAs, GaP, GaSb, GaN, InAs, InP, $Ga_xIn_{1-x}P$, $InAs_yP_{1-y}$, $GaAs_yP_{1-y}$, $GaAs_{1-y}Sb_x$, and $Ga_{1-x}In_xAs_yP_{1-y}$ [56]. Using appropriate liners, Al-containing compounds such as, for instance, AlAs, AlP, AlN, and $Ga_xAl_{1-x}As$, have also been grown

[56]. The MOCVD technique is, however, much superior [41] for growing Al-containing compounds. Our discussion will focus on the growth of GaAs, InP, Ga_xIn_xAs, and $Ga_xIn_{1-x}As_yP_{1-y}$ that are used in or have potential for microwave and gigabit-rate technology for communications applications.

We will briefly describe the GaAs growth system, and insert appropriate comments about InP and GaInAsP growth. Figure III-4(a) is a schematic diagram of the GaAs reactor; figure III-4(b) is a schematic for the growth of GaInAsP alloys. A recent review of the growth of GaInAsP alloys for electro-optic and microwave applications will be found in references 57 and 58, respectively.

The key features of the hydride process are as follows:

(1) Reaction of the group III metal with high-purity HCl and then transporting it as the chloride. Independent metal sources (e.g., Ga and In) are provided to grow alloys. The typical source temperature is 850°C.

(2) The group V material is introduced downstream of the group III metal as the hydride (e.g., AsH_3, PH_3). These hydrides decompose and react with the group III chloride; the III-V compound is deposited on the substrate typically at 700°C. The chemistry of the CVD processes occurring during synthesis has been extensively studied by Ban [59] by coupling a time-of-flight mass spectrometer to a CVD reactor.

(3) The n-type doping can be controlled by the use of dopants such as S, Se, or Si (usually introduced with the group V hydride) in the form of a gaseous hydride such as H_2S, H_2Se, or SiH_4.

(4) p-type doping can be carried out with Zn. The comments made for the trichloride process are also applicable to the hydride system.

iv. MOCVD Process: The first demonstration of the growth of GaAs by MOCVD was by Manasevit and Simpson in 1969 [32]. Interest was rekindled in this process when Bass demonstrated the growth of material of microwave device quality [60]. Since then many high-performance devices have been realized by the use of MOCVD-grown material including solar cells, lasers, quantum well heterostructures, photocathodes, and GaAs low-noise and power FETs. A special issue of the Journal of Crystal Growth [61] describes these efforts in detail. The following characteristics describe an MOCVD system:

(1) Group III metals are transported by metal alkyl vapor, e.g., trimethylgallium (TMG), TMAl, TMIn, etc.; group V compounds are transported as the hydride (e.g., AsH_3, PH_3).

AsH_3+H_2
H_2Se+H_2 (DOPANT)
H_2
SUBSTRATE
$HCl+H_2$
Ga
CONTINUES TO EXHAUST
GALLIUM ZONE
REACTION ZONE
DEPOSITION ZONE

(a) Vapor hydride apparatus for GaAs growth.

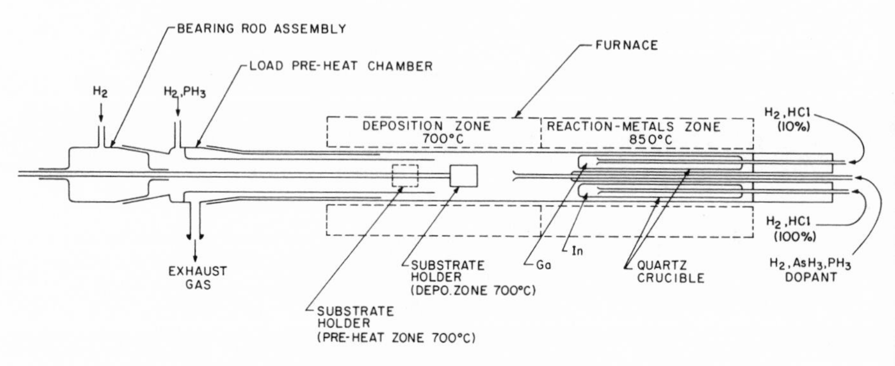

(b) Quaternary reactor tube.

Figure III-4. Schematic representation.

(2) The basic reaction is an irreversible pyrolysis that takes place on the heated substrate analogous to the pyrolysis of SiH_4 in Si epitaxy. Since only the substrate is heated, a cold-wall system with rf or infrared heating can be used. This makes for simpler equipment than necessary for the halide transport methods and is easier to scale up for production.

(3) High gas flows can be used since there is no source material that has to be equilibrated. This fact, plus the absence of halide species, limits autodoping.

(4) GaAs layers can be grown at temperatures as low as 600°C. This minimizes diffusion from the substrate.

(5) As indicated earlier, Al-containing compounds can be grown.

(6) Epitaxial growth at reduced pressure (∿76 Torr) can be achieved. Low-pressure growth results in lower autodoping.

High-purity GaAs with total ionized impurity concentration ($N_D + N_A$) of about 5×10^{14} cm^{-3} with liquid-nitrogen mobility (μ_{77}) of 125-135x10^3 cm^2 V^{-1} s^{-1} has been realized with MOCVD GaAs [62]. Low-noise GaAs FETs with an excellent noise figure have also been reported [63]. GaAs power FETs fabricated from MOCVD GaAs have demonstrated performances comparable with those of FETs with active layers grown by halide epitaxy [64]. The dominant residual impurities are C, Si, and Zn [62]. C is the only impurity inherent to the process, and indications are that the use of triethylgallium (TEG) in conjunction with low-pressure epitaxy can reduce C incorporation.

The disadvantages of the MOCVD process, for the most part, are not inherent to the process and are due to the current state of the art. These are the disadvantages:

(1) The purity of the grown layers is a strong function of the metal alkyl purity. The purity of metal alkyls is not consistent and varies from batch to batch and vendor to vendor. This is not an inherent disadvantage. The metal alkyl purity is, however, continuously improving as demand increases.

(2) Growth of InP and GaInAs is complicated by the fact that In alkyls and PH_3 undergo parasitic reactions and form addition compounds. This can be compensated for by precracking the PH_3 or by the use of trimethyl PH_3 (organohydride) or coordination compounds. A lot of research is being done in this area.

While most of the devices currently use material grown by chloride epitaxy, MOCVD is making some inroads. The major attraction is the ease of scaleup and growth of Al-containing compounds. Some researchers claim that barrel reactors employing low-pressure CVD operating around 600°C can compete effectively with ion implantation, especially for discrete microwave device applications [64].

b. Molecular-Beam Epitaxy (MBE)

Molecular-beam epitaxy is a process of epitaxial deposition using molecular or atomic beams in an ultrahigh vacuum (UHV ∿ 10^{-10} Torr) system. In principle, MBE is a technique of vacuum evaporation, one of the oldest methods for depositing thin solid films. The molecular or atomic beams are usually generated thermally in Knudsen cells where quasi-equilibrium is maintained. The beam compositions and intensity are therefore constant and predictable from thermodynamics. The beam fluxes, guided and controlled by orifices and shutters, travel in straight paths to the substrate; there they condense and grow under kinetically controlled conditions. Since the MBE process takes place in a UHV environment, many in situ analytical measurements can be made to shed light on the growth. Accessory equipment that can be incorporated in an MBE system includes mass spectrometers, Auger analyzers, ion-bombardment apparatus, and a variety of electron microscopy and diffraction systems. The selection of the analytic equipment depends on the specific objective of the investigation. MBE systems have been used for the growth of group IV, III-V, V, II-VI, and IV-VI compounds. An excellent tutorial article with a considerable number of references is available in Vol. III of the Handbook on Semiconductors [39]. Table III-5 lists some of the many compounds grown by MBE.

Table III-5. SEMICONDUCTOR GROWN BY MBE [67]

IV	III-V	II-VI	IV-VI
Si	GaAs	ZnTe	PbTe
Ge	(Ga,Al)As	ZnSe	(Pb,Sn)Te
SiGe	Ga(As,P)	Zn(Se,Te)	PbS
	InP	CdTe	PbSe
	GaP	CdS	SnTe
	(In,Ga)As		(Pb,Sn)Se
	Ga(Sb,As)		
	AlAs		

The MBE process possesses some unique features vis à vis various competitive techniques. Among these features are low deposition temperature (500-600°C), ability to achieve atomically smooth surfaces, flexibility in incorporating a number of source beams to grow compounds with differing composition, growth of ultrathin layers, complex profiles in terms of composition and concentration, and the growth of sophisticated structures by successive deposition without thermal diffusion and redistribution. These features are exemplified by the achievement of periodic structures on the scale of atomic dimensions, e.g., modulation-doped heterojunction superlattices [65].

These advantages must be counterbalanced against the relatively high initial cost and the current low throughput. For discrete low-noise and medium-power field-effect transistors made from a single semiconductor or a simple heterostructure, other epitaxial techniques may be adequate and more cost-effective. For monolithic microwave and gigabit-rate GaAs ICs, the ion-implantation technique may be more suitable. Devices and circuits using single-period modulation-doped heterostructures (e.g., high-mobility FETs [66]) may require MBE (or MOCVD). While it is too early to judge the commercial utility of the MBE process, the developments of the next two to three years may allow such an assessment. Several commercial MBE systems are now being marketed.

To illustrate the process, we briefly describe an MBE system for the growth of GaAs [67]. A typical system is shown in figure III-5. The system is first pumped by standard mechanical and sorption pumps and then by a combination of ion and titanium sublimation pumps. The entire system is usually bakeable to temperatures of about 250°C and reaches ultimate pressures of the order of 10^{-10} Torr. Residual gas species typically include CO, H_2O, H_2, CH_4, and CO_2 in decreasing order of abundance. The dominant impurity, CO, is difficult to eliminate. The source ovens are of pyrolytic BN or high-purity graphite and are resistively heated (e-beam heating is sometimes employed; e.g., for Si). All ovens are surrounded by a liquid-nitrogen shroud with appropriate thermocouples, control orifices, and shutters. The temperatures of the individual furnace are chosen so that the vapor pressures of the materials are high enough for the generation of thermal-energy molecular beams. The furnaces are arranged so that the central portion of the beam fluxes impinge on the heated substrate.

The bakeout and pumping procedures for attaining UHV are too time consuming and tedious to allow exposure of the vacuum system to air for sample loading and unloading. Specimen exchange load-locks are therefore used, permitting operation over long time periods (weeks to months) with the vacuum unbroken. Properly designed radiation heat shields, baffles, etc. are required to prevent thermal and chemical cross-contamination. A brief tutorial description of MBE equipment can be found in reference 67.

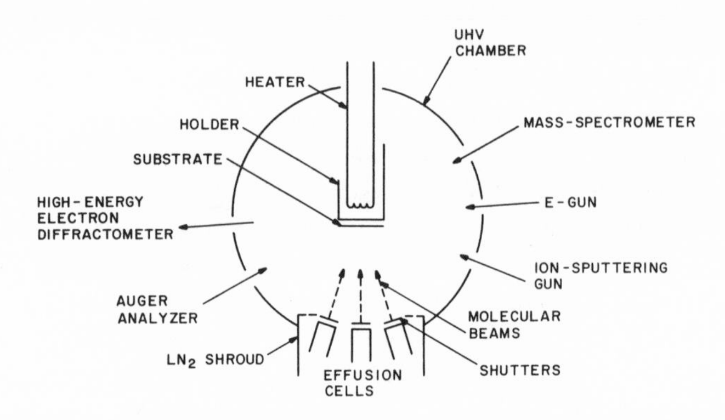

Figure III-5. Schematic diagram of a molecular-beam epitaxy system.

The substrate is generally mounted with indium to a uniformly heated block of refractory material such as molybdenum or BN. Temperature uniformity is of obvious importance. Indium, liquid at typical growth temperatures, holds the substrate to the block by surface tension and provides good heat transfer. Ga and Sn have also been used.

The substrate can be rotated from the growth to various analysis and/or cleaning positions without change in temperature. In situ cleaning techniques include heating under UHV, heating in 10^{-6} Torr ambient of O_2 or H_2O, and sputtering with low-energy (500 eV) ions, followed by annealing to remove ion damage. Table III-6 lists surface analytical techniques most applicable to MBE and the typical uses they are put to. In principle, once the growth process is developed, the analytical equipment can be eliminated, thus reducing the cost of production equipment significantly. This contention remains to be firmly established.

A characteristic of the MBE process is the low growth rate. Typical growth rates are micrometer-per-hour or approximately a monoatomic layer per second. The beam fluxes can thus be modulated in monolayer quantities, resulting in the growth of very thin films with abrupt interfaces, provided the substrate surface is atomically smooth and the growth temperature is low enough for negligible diffusion. This low growth rate necessitates a UHV environment, or contaminants with reasonably large sticking coefficients will be incorporate

Table III-6. IN SITU ANALYTICAL TECHNIQUES [67]

Technique	Functions
Mass Spectroscopy	1. Molecular-beam flux analysis 2. Residual-gas analysis
Reflection High-Energy Electron Diffraction (RHEED)	1. Surface crystal structure 2. Surface roughness
Auger Spectroscopy	1. Surface analysis (<5-μm spatial resolution): Stoichiometry Chemical states Contamination 2. Depth profiles (∿50-Å resolution): Matrix composition profiles Dopant profiles in heavily doped material
SIMS	1. Surface analysis (>100-μm resolution): Contamination Chemical states 2. Depth profiles (∿50-Å resolution): Matrix composition profile Dopant profiles - very sensitive for certain elements

in the grown layer. A dramatic example of multilayers that can be achieved by MBE is the growth of 10^4 alternating GaAs and AlAs monolayers [65].

For compound semiconductors in which congruent evaporation does not occur at reasonable temperatures, it is not obvious that stoichiometric growth is possible. Furthermore, growth is determined by complex kinetic reactions at the substrate surface; even congruent molecular beams need not result in stoichiometry. Fundamental studies of absorption-desorption kinetics [68] have shown that the solution to the stoichiometry problem in III-V compounds lies in the surface chemical dependence of the sticking coefficient of the group V elements. At epitaxial growth substrate temperatures (>500°C), group V compounds are adsorbed only until the available group III orbitals at the surface are satisifed. Growth rate, in general, depends on the arrival rate of the group III elements, while stoichiometry is maintained by allowing an excess group V flux. It has been proposed that a step growth mechanism exists at the substrate surface. The group III and V atoms are initially bound in mobile precursor states and migrate till step edges (e.g., facets) are encountered. Incorporation into growing film occurs at these edges, and steps propagate until they coalesce. Step densities are determined by the microscopic flatness of the substrate, which improves as the epitaxial film grows. These densities have been ascertained by Pt-C replication electron microscopy [69]. This oversimplified discussion illustrates the complexity of the process.

The incorporation of dopants into the matrix semiconductor is also a complex process. Anomalously high diffusion coefficients, surface-chemistry-dependent sticking coefficients and lattice site incorporation, and surface segregation have been reported for dopants. However, order-of-magnitude changes of about 50 Å have been reported for some dopants [67]. Table III-7 shows some common dopants reported in the literature.

In summary, MBE has not yet been developed into a production process. It is particularly suited to unique multisemiconductor structures like modulation-doped heterostructures. The potential application areas are in optoelectronic devices (most current research appears devoted to this area) and specialized microwave and gigabit-rate logic devices.

c. Liquid-Phase Epitaxy (LPE)

The deposition of semiconductor films by liquid-phase epitaxy (LPE) has been developed into a very useful technique for the preparation of many III-V compounds and alloys. Some of the purest GaAs, InP, and GaInAs grown have been produced by LPE. This technique is also useful for growing Al-containing compounds [34]. LPE requires relatively simple apparatus compared to VPE and MBE, has generally higher growth rates, and offers a large selection of available dopants. LPE systems are in general more difficult to scale up and have lower throughput than VPE systems. Currently, LPE is used more for optoelectronic than microwave devices, due especially to the ease with which GaAlAs/GaAs heterostructures can be grown. Good surface morphology, on the other hand, is easier to obtain with VPE than LPE. This is of paramount importance if submicrometer lithography is required.

Since LPE is a well-known and relatively mature technique, extensively described in the literature [34] and now only sparingly used in most microwave

Table III-7. PROMISING DOPANTS* [67]

n-Type

Sn - Doping concentration to 1.2×10^{19} cm^{-3}. Low compensation for concentrations to 1×10^{18} cm^{-3}.

Tends to segregate at growth surface, blunting sharp profiles, particularly at elevated temperatures.

Excellent photoluminescent efficiency for Sn-doped GaAs.

Most commonly used n-type dopant.

Si - Doping concentrations to 5×10^{18} cm^{-3}.

Slightly more compensation than for Sn.

Negligible diffusion and segregation permits sharp profiles.

Excellent photoluminescent efficiency for Si-doped GaAs.

p-Type

Mn - Doping concentrations to 1×10^{18} cm^{-3}.

Deep acceptor (∿113 meV above valence band edge).

Surface stiochiometry influences dopant incorporation.

Mg - Electrical activity strongly dependent on Al concentration in $Al_xGa_{1-x}As$. Acceptor concentration of 2×10^{16} and 1×10^{19} cm^{-3} obtained for $x = 0$ and $x = 0.2$, respectively.

Be - Doping concentrations to 3×10^{19} cm^{-3}.

Shallow acceptor (30-35 meV above valence band edge).

Dopant incorporation independent of substrate temperature and Al concentration.

Negligible diffusion or segregation permits sharp profiles.

Zn - Zn doping concentrations to 10^{19} cm^{-3}.

Beam flux easily monitored by ion current to substrate.

*Examples only; not exhaustive.

and gigabit-logic applications, our discussion will be very brief. Reference 39 contains a tutorial discussion with many references to the literature.

LPE is basically the precipitation of a crystalline layer from the liquid phase onto a parent substrate. The crystallographic orientation of the grown layer is determined by that of the substrate. Detailed analysis of the LPE process is a difficult undertaking that involves consideration of many factors, including phase equilibria, nucleation, interface kinetics, interface morphology, solute partitioning, defect generation, and thermal transport. Examples of III-V compounds grown by LPE include GaAs, InP, GaAlAs and GaAs/GaAlAs heterojunctions, GaInP/InP, and GaInAsP/InP.

d. Comparison of Epitaxial-Growth Technologies

Table III-8 summarizes the key characteristics of the various epitaxial technologies for the growth of III-V compounds.

3. Ion Implantation

Ion implantation is a very powerful technique for the uniform, controlled, and selective doping of semiconductors. Ion implantation is now a widely used production technique for Si and is rapidly becoming very widely used for III-V compounds, particularly GaAs. The general aspects of ion implantation have been covered in many reviews [39]; the ion implantation into GaAs has been reviewed by Donnelly [70]. We will present a brief summary of the more important considerations with emphasis on GaAs.

An energetic ion entering a thick, solid target will lose its energy in a series of collisions with the nuclei and electrons in the target and will finally come to rest. In amorphous targets, the typical distribution of the implanted species is almost Gaussian and can be described by an average range R_p and a standard deviation (also called straggle) ΔR_p, given by the equation [70]

$$N(x) = \frac{N_s}{(2\pi)^{2/3}\,(\Delta R_p)} \exp\left[-(x - R_p)^2/2(\Delta R_p)^2\right] \qquad (1)$$

where N_s is the dose or fluence (atoms/cm^2) and x is the perpendicular distance into the substrate. The projected range and standard deviation of many ions into GaAs have been tabulated [71]. For some ion/substrate combinations,

Table III-8. COMPARISON OF EPITAXIAL-GROWTH TECHNOLOGIES

Technique	Factors Controlling Solid Composition	Factors Influencing Layer Purity	Advantages	Disadvantages	Current Status
1. Liquid-Phase Epitaxy (LPE)	Phase diagram, interface kinetics, thermal transport	Group III melt, container, gettering, bakeout time and temperature	Simple apparatus, high purity	Difficult to scale up; limited throughput; morphology difficult to control	Small-scale systems for opto-electronics laboratory technique
2. Vapor-Phase Epitaxy (VPE)					
a. Hydride Process	Thermodynamics	Gas purity, reactor materials, leaks	Flexible - control over gas phase composition (i.e., III/V ratio); large-volume reactors possible	Al-containing alloys difficult to grow; lower purity than LPE	Production technique for GaAs, GaAsP
b. Trichloride Process	Thermodynamics	Gas purity, reactor materials, leaks	High-purity binaries; large-volume binary reactors feasible	Al alloys difficult; less flexible than the hydride - no control over III/V gas phase ratio	Most popular production technique for GaAs microwave device applications
c. MOCVD Process	Kinetics - arrival rate at surface	Metal-oxide-source purity, hydride purity, C-contaminants, O_2 and H_2O traces	Simple apparatus, easy to scale up for production; Al-containing compounds can be grown	C contamination; parasitic reactions affect In- and P-containing compounds	Laboratory; significant interest in scaling up
3. Molecular-Beam Epitaxy (MBE)	Kinetics - beam flux, sticking coefficient	Vacuum system, UHV pressure, residual gases	Relatively low temperature (∿500-600°C), very abrupt interfaces, modulation-doped heterostructures with very abrupt (∿20 Å) interfaces, in situ analysis	Expensive, low growth rate, problems with P-containing alloys; production process may be difficult	Special opto-electronics; microwave and gigabit-rate logic structures

considerable skewing of the implanted profile occurs and higher-order moments have to be included. For implantations made through masks, lateral spreading at mask edges must also be considered.

In crystalline targets, the implanted distribution depends on the relative orientation of the beam to major crystal axes. If ions are implanted parallel to a major axis, considerably higher ranges are obtained due to "channelling." To prevent channelling, the crystal axes are usually inclined away from the beam (typically 5-7°). Since some ions are deflected into channelling directions, tails usually occur on otherwise Gaussian distributions. While the energy of ions usually implanted into GaAs ranges from 20 to 400 keV, implantation up to 1.2 MeV has been reported [72].

The implanted ions interact with the target nuclei, and, as they come to rest, displace some nuclei and even cause a cascade of displacements. The amount of damage produced depends upon ion mass, dose, and temperature of implantation. Furthermore, the implanted atoms, in general, do not occupy substitutional sites. The implant damage must therefore be annealed out and the implanted ions forced to occupy the requisite substitutional site. Annealing is achieved by heating in a furnace for a specific period of time, by high-power laser irradiation [73], electron bombardment [74], or exposure to incoherent radiation from an arc or quartz-halogen lamp [75]. Sometimes, during annealing, implanted species may diffuse at extraordinary rates due to damage-assisted diffusion, and impurity redistribution in the substrate may occur.

A major complication in the annealing of ion-implanted compound semiconductors lies in the fact that one of the components (usually group V elements) tends to preferentially evaporate at temperatures below about 600°C. Annealing temperatures, however, lie in the 600-900°C range. As a result, it is generally necessary to encapsulate the material before annealing. Various dielectric materials such as SiO_2, Si_3N_4, Al_2O_3, AlN, and metals such as Al, have been used as encapsulants. Annealing without encapsulation (capless) in a controlled atmosphere has also been reported [72]. Capless annealing, when possible, is preferred since problems due to diffusion of a constituent from the encapsulation and diffusion into the encapsulant can be avoided.

The major advantages of ion implantation for device fabrication are as follows:

(1) Ion implantation provides very precise control over the dopant penetration depth by control over implant energy.

(2) The total number of ions implanted depends upon the time-integrated beam current that can be accurately controlled.

(3) By rastering the ion beam over the target many times during the implantation, excellent lateral doping unformity can be achieved.

(4) By using photolithographically delineated implant masks, selective doping over a substrate can be achieved.

(5) Multiple energy/dose implant schedules allow realization of fairly complex doping profiles perpendicular to the substrate surface.

(6) Implant-induced damage can be utilized to render areas of semiconductor wafers insulating [70] to provide isolation.

(7) Many ions can be implanted without regard to chemical barriers or surface effects that might otherwise inhibit the introduction of the species.

The disadvantages are residual defects left after annealing, damage-enhanced diffusion, and in the case of SI GaAs, impurity redistribution at the implanted-layer substrate interface. The advantages far outweigh the disadvantages.

We now briefly consider ion implantation into GaAs, in particular SI GaAs, since this is of paramount importance for GaAs gigabit-rate digital ICs, monolithic microwave integrated circuits, and discrete microwave devices such as MESFETs. An excellent review is presented in reference 70. The implantation of donors up to energies as high as 1.2 MeV to obtain 1-μm-thick n layers in SI GaAs has been described by Liu et al. [73]. These authors also describe an operationally simple, capless anneal process under As over-pressure that results in high activation with excellent surface morphology and minimum substrate impurity redistribution [73].

As discussed in the section on SI GaAs substrates, the quality of the implanted layers is a strong function of substrate characteristics [3]. Substrate quality affects the implanted layers' electron mobility, mobility profile, doping profile, activation efficiency, run-to-run reproducibility, substrate impurity distribution, and conversion of the insulating to conducting characteristics [2,3]. The method of annealing also has a strong effect on impurity redistribution [73]. In general, capless annealing gives far better results than does capped annealing. With the continuous improvement in SI GaAs substrate quality, more and more laboratories are moving toward the use of ion implantation. It is still necessary to "qualify" SI GaAs slices from every substrate boule by test implantation in samples out from the seed and tang ends One can compensate for substrate quality somewhat by growing a high-resistivity epitaxial "buffer" layer; this is not, however, a long-term solution.

The current state of the art of direct ion implantation into SI GaAs substrates can be exemplified by the fact that LSI circuits with 1000-gate complexity (8x8 multiplier) have been achieved [76]. GaAs power FETs operating at frequencies as high as 26 GHz have also been reported [77]. Table III-9 lists some common n- and p-type species that have been implanted into GaAs. Figure III-6 shows the characteristics of a 1-μm-thick layer generated by ion implantation [73].

The implantation into InP and other III-V compounds and alloys is still in a rudimentary state. The same technological considerations, however, apply.

A major factor influencing the development of ion-implantation techniques is that it is widely used in Si technology. A result of this wide technology base is that the equipment is constantly improving and that there is a large effort to understand the basic physics of the process. These factors leverage the development of implantation and annealing techniques for GaAs. Some examples of this effect are laser annealing [73], electron-beam annealing [74], and annealing by means of radiation from an incoherent quartz-halogen lamp. These techniques are not yet developed to the extent of their being used in actual device fabrication. Their major potential advantage is that they are "transient" and capable of being flexibly incorporated into the device fabrication schedule. Usually furnace annealing must be carried out before any device metallization is applied to prevent excessive diffusion. Transient annealing, in principle, can be done after metallization, thus offering more flexibility in incorporating implantation into the device fabrication schedule.

4. Current Status and Trends

The current status and trends in III-V epitaxial technology and ion implantation are described in table III-10.

Table III-9. SUMMARY OF ELECTRICAL CHARACTERISTICS OF GaAs IMPLANTED WITH n- AND p-TYPE IMPURITIES [70]

n-Type

1. Temperature of Implant		2. Anneal Temperature (typical)
Si	room temperature	800-900°C
Sn	>100°C (high dose)	900°C
S	>150°C (low dose - room temperature)	800-900°C
Se	>150°C	900°C
Te	>150°C	900°C

p-Type

Beryllium

1. Implant temperature - room temperature
2. Anneal temperature
   a. High-temperature anneal: 900°C

      Low dose - implanted Be $<5x10^{18}$ cm^{-3}
      ∿100% electrical activity
      No diffusion

      For p ∿ $2x10^{18}$ cm^{-3}, μ = 120 cm^2 V^{-1} s^{-1}

      High dose - implanted Be > $5x10^{18}$ cm^{-3}
      Diffusion - flat profiles

      For p ∿ $5x10^{18}$ cm^{-3}, μ ∿ 120 cm^2 V^{-1} s^{-1}

   b. Low-temperature anneal: 650°C
      (Note: this low-temperature activation is now always observed)

      ∿100% electrical activity
      No diffusion

Cadmium and Zinc

1. Room temperature or hot implants (T > 150°C)
2. 800-900°C anneals
3. Diffusion (dose-dependent)
4. ∿100% electrical activity for Zn or Cd < 10^{20} cm^{-3}
5. A peak of p $\gtrsim$ $2x10^{19}$ cm^{-3} can be achieved
6. Hot implant decreases diffusion
7. Dual implant (Cd + As) decreases diffusion

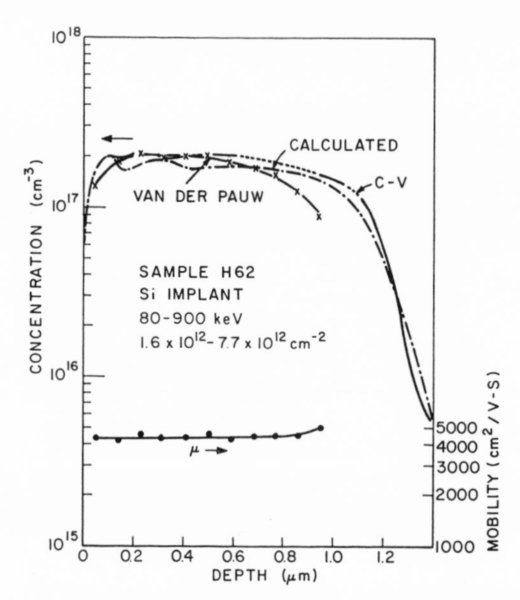

Figure III-6. Multiple-implant profiles in SI GaAs:atomic profile by SIMS, carrier concentration, calculated profile, and mobility profile shown.

B. MICROWAVE SOLID-STATE DISCRETE ACTIVE DEVICES

The discovery of the Gunn effect in 1963 marks the beginning of the development of compound semiconductor microwave devices. Dramatic progress in the last two decades has led to the application of compound semiconductor microwave devices, especially GaAs devices, in many of today's microwave systems. Communications, radar, and EW systems currently in use offer advantages in performance, size, reliability, and cost that would have been impossible without these devices.

One particular event that has revolutionized the field of III-V semiconductor microwave devices is the development of the GaAs field-effect transistor. This versatile device has found application in low-noise amplifiers, medium-power amplifiers, and oscillators. Dual-gate GaAs FETs are being employed as

Table III-10. CURRENT STATUS AND TRENDS

Technology	Current Status	Future Trends
1. LPE	1. Growth of GaAs/GaAlAs structures for commercial lasers and other optoelectronic devices	1. Competition from MOCVD and MBE may make inroads
	2. Growth of (Ga,In)(As,P) alloys for optoelectronic devices	2. MOCVD and ion implantation may supplant LPE for microwave device and gigabit-rate logic applications
	3. Some GaAs being grown for microwave and gigabit-rate logic applications	
2. VPE		
a. Chloride Process	1. Dominant process for growth of material for discrete microwave low-noise and power FETs	More automated, large-diameter, multiwafer reactors for GaAs and InP growth
	2. Some multiwafer reactors with 60- to 75-mm-diam wafer growth capability	
	3. Commercial reactors available	
b. Hydride Process	1. Production reactors for GaAsP LEDs	More automated, large-diameter, multiwafer production reactors for GaAs, InP, and (Ga,In)(As,P) growth
	2. Strong second for growth of GaAs for microwave and gigabit-rate logic applications	
	3. Commercial multiwafer reactors, 60- to 75-mm-diam substrate capability available	
	4. Laboratory reactors for growth of (Ga,In)(As,P) alloys for optoelectronics and for emerging microwave and gigabit-rate logic applications	

Table III-10. (Continued)

Technology	Current Status	Future Trends
c. MOCVD Process	1. Laboratory process, demonstrated growth of GaAs, GaAlAs, and GaAlAs/GaAs heterostructures 2. Low-pressure MOCVD-grown GaAs with very abrupt interfaces demonstrated 3. GaAs FETs fabricated from MOCVD GaAs demonstrated very low noise figures 4. Purity limited by that of organometallic sources 5. Some problems in growing In- and P-containing compounds due to parasitic reactions 6. Some commercial reactors becoming available	1. Commercial low-pressure, barrel-type MOCVD systems with 60- to 75-mm diam, multiwafer (~25) will become available for growing GaAs/GaAlAs 2. Development of above reactors may supplant chloride/hydride reactors for GaAs 3. May prove to be the cost-effective alternative to the growth of GaAs/GaAlAs heterostructures for fabricating the following: a. Modulation-doped FETs for microwave and multigigabit-rate logic applications b. Wide-bandgap emitter bipolars for microwave and multigigabit-rate logic c. Optoelectronic applications
3. MBE	1. Laboratory process for growing special structures such as heterojunctions or quantum well structures 2. In situ analytical instrumentation allows fundamental growth kinetics studies	1. If MOCVD cannot grow quality heterojunctions, MBE may be the only alternative for modulation-doped structures 2. While production systems with reduced analytical capabilities will become available, equipment is expected to be more expensive than for other epitaxial technologies 3. Will find use for growing special structures

Table III-10. (Continued)

Technology	Current Status	Future Trends
4. Ion Implantation (Doping - not matrix growth)	1. Implantation of donors up to 1 MeV demonstrated; the energies most used are 20-400 keV; a 400-keV Si implant results in 0.35- to 0.4-μm-thick n layer in GaAs	1. Will find application in GaAs microwave and gigabit-rate logic as SI GaAs substrate quality continues to improve
	2. Implantation of acceptors in GaAs up to 400 keV demonstrated	2. Radiant annealing by means of pulsed (3-10 s) quartz halogen or arc lamps will make process more cost-effective; radiant anneal will also result in more flexible integration of implantation with device processing
	3. Mature production technique for Si devices; equipment handling 100-mm-diam substrates, multisubstrate end stations, and 400-keV machines are commercially available	3. Will probably be the dominant doping technology in GaAs FET-based monolithic microwave technology and multigigabit-rate ICs
	4. Operational capless and encapsulated furnace-annealing techniques developed	4. More effort on implantation into InP and GaInAs/InP alloys
	5. Laser, e-beam, and radiant-heat annealing for GaAs being developed	
	6. Techniques for implant/anneal of donors and acceptors in InP and other III-V compounds being developed; qualitatively similar to GaAs	

mixers, switches, amplitude-controlled amplifiers, limiters and limiting amplifiers, and frequency discriminators. Furthermore, single- and dual-gate GaAs FETs are key elements in monolithic microwave and multigigabit-rate digital integrated circuits. Frequency coverage of GaAs FETs is continuously expanding and has now reached the 26- to 40-GHz band.

In addition to the maturing GaAs FET technology, many novel device configurations of potential importance are being studied in many laboratories all over the world. Current research includes the investigation of other III-V compounds, such as $Ga_{0.47}In_{0.53}As$, that have electrical properties superior to those of GaAs; and of devices such as wide-bandgap-emitter bipolar transistors and permeable-base transistors.

This section concentrates on discrete devices. Table III-11 summarizes the currently important discrete III-V compound active microwave devices. Table III-11 is divided into commercially available devices, devices under advanced development, and exploratory devices. We will consider all of these categories, with emphasis on the latter two.

We have divided solid-state active microwave devices into two categories, viz., two- and three-terminal devices. The characteristic features of two- and three-terminal active microwave devices are as follows:

(1) Two-terminal active microwave devices are negative-resistance devices. They are easier to use in oscillator than in amplifier applications. A circulator or a 90° 3-dB hybrid must be used to separate input and output signals when amplification is required. Negative-resistance devices can serve as locked oscillators for high-gain, relatively narrowband amplification or as stable reflection amplifiers. The negative resistance and reactance of two-terminal devices are a function of drive level, leading to problems in linear communications amplifier applications. Gewartowski describes a case history of the design of an IMPATT amplifier for a Bell System repeater application [78] that illustrates the many problems involved in amplifier design with two-terminal oscillators.

In contrast, three-terminal devices, such as transistors, possess excellent input/output isolation, are unconditionally stable over large bandwidths, are easy to characterize under small- and large-signal conditions, and are well suited to amplifier applications. With external feedback networks, they can be used as oscillators.

Table III-11. DISCRETE III-V COMPOUND ACTIVE MICROWAVE DEVICES

Commercially Available	Advanced Development	Exploratory Development
Two-Terminal Devices	Two-Terminal Devices	Three-Terminal Devices
• GaAs TEDs (Gunn devices) 50 GHz - 100 mW 94 GHz - 20 mW 110 GHz - 2-5 mW	• InP TEDs	• GaAs MESFETs 18-40 GHz • InP MISFETS • $Ga_{0.47}In_{0.53}As$ MISFETs
• GaAs IMPATTs 15 GHz - 2 W 20 GHz - 1 W	Three-Terminal Devices • Low-Noise GaAs MESFETs 12- to 26-GHz Band	• GaAs Permeable-Base Transistors (PBTs) - f_T >100 GHz
Three-Terminal Devices • Low-Noise GaAs MESFETs 4 GHz - 1 dB NF 12 GHz - 2.5 dB NF	• GaAs Power MESFETs S-Band - 5-20 W X-Band - 1-5 W Ku-Band - 1-2 W	• GaAs/GaAlAs Modulation-Doped Channel FETs • Ballistic Submicrometer Channel Devices
• GaAs Power MESFETs S-Band - 1-5 W X-Band - 1 W Ku-Band - 0.5 W		• GaAs/GaAlAs Heterojunction Bipolar Transistors (HBTs)

(2) Two-terminal active devices utilize the transit time of carriers through the device to generate the negative resistance. In contrast, the frequency response of three-terminal devices is limited by transit time effects. Two-terminal devices are currently easier to fabricate than three-terminal devices for frequencies of 20 GHz and higher. For example, GaAs TEDs and Si IMPATTs operate at frequencies up to 100-120 GHz and 200-250 GHz, respectively.

In summary, two-terminal devices are particularly suited for oscillator applications (e.g., local oscillators in receivers, beacon transmitters, VCOs, etc.) and at frequencies where three-terminal devices are not available. At frequency regions where both two- and three-terminal devices are available, as a general rule the three-terminal device is preferred. Table III-12 summarizes the qualitative advantages of three-terminal devices for amplifier applications.

Table III-12. ADVANTAGES OF THREE-TERMINAL OVER TWO-TERMINAL DEVICES FOR AMPLIFIERS

- Excellent Input/Output Isolation
- Well-Behaved Amplifier Characteristics
  - Bandwidth Relatively Insensitive to Drive
  - Monotonic Gain Saturation
  - Straightforward Small- and Large-Signal Characterization
- Good Linear Amplifier
  - Low Intermodulation Distortion
  - Low Variation of Phase Shift with Drive
  - Lower AM/PM Conversion
  - Better Phase Linearity
- Much Lower Noise Figure
- Simpler Circuits
- Easier Temperature Compensation
- Smaller Amplifier Size and Weight

1. Two-Terminal Devices

Active two-terminal microwave devices are negative-resistance devices. The real part of their impedance is negative over a range of frequencies. In a negative-resistance element, the current and voltage are 180° out of phase, and a current I flowing through a negative resistance -R will produce a voltage rise -IR, while a power of $P = I^2R$ will be generated by the power supply associated with the negative resistance. Figure III-7 is a plot of the real and imaginary parts of the impedance Z associated with a typical negative-resistance device. In this example, Re[Z] < 0 over the range from 8 to 16 GHz.

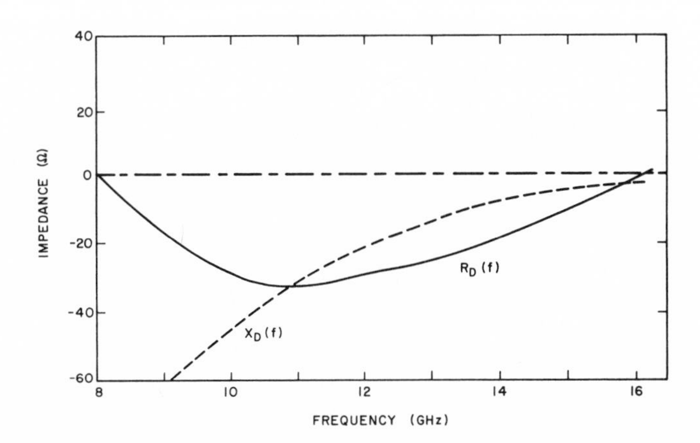

Figure III-7. Real and imaginary parts of the impedance of a typical transferred-electron device vs frequency.

In microwave applications, the active device is mounted at the end of a transmission line. If the real part of the circuit impedance, $Re(Z_c)$, is negative at a given frequency, then an rf signal incident upon the circuit will be amplified and the power in the reflected wave will be larger than that in the incident wave. The reflection gain, G, defined as the ratio of the reflected power R_R to the incident power P_{in} is given by

$$G = \frac{P_R}{P_{in}} = \frac{\left[R_o - R_e(Z_c)\right]^2 + \left[I_m(Z_c)\right]^2}{\left[R_o + R_e(Z_c)\right]^2 + \left[I_m(Z_c)\right]^2} \qquad (2)$$

where R_o is the characteristic impedance of the transmission line. Clearly when $R_e(Z_c) > 0$, the power in the reflected wave is smaller than that in the incident wave. In practical reflection amplifiers, which contain negative-resistance elements, the input and output signals are separated by a ferrite circulator or a 90° 3-dB hybrid, as shown in figure III-8.

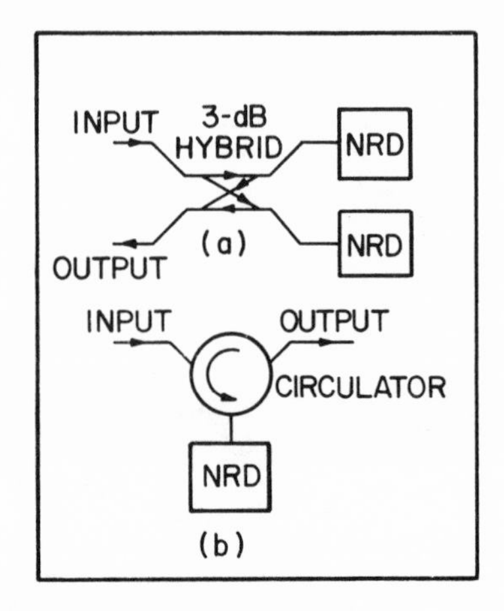

Figure III-8. Reflection amplifier using two-terminal negative-resistance devices.

In negative-resistance oscillators, a resonant circuit and load are placed across the negative-resistance element, as shown in figure III-9. The system will break into spontaneous oscillation if there is a net negative resistance at the resonant frequency. In general, two-terminal negative-resistance devices are easier to use as oscillators than amplifiers.

The two-terminal negative-resistance microwave devices of current interest are as follows:

i. Transferred-electron (Gunn) devices: In transferred-electron devices (TEDs), negative resistance is achieved by taking advantage of the negative differential mobility (dv/dE) of electrons in certain n-type III-V compounds, mainly GaAs and InP. The basic theoretical concepts underlying the operation of TEDs were developed during the early 1960s; the first experimental TEDs were

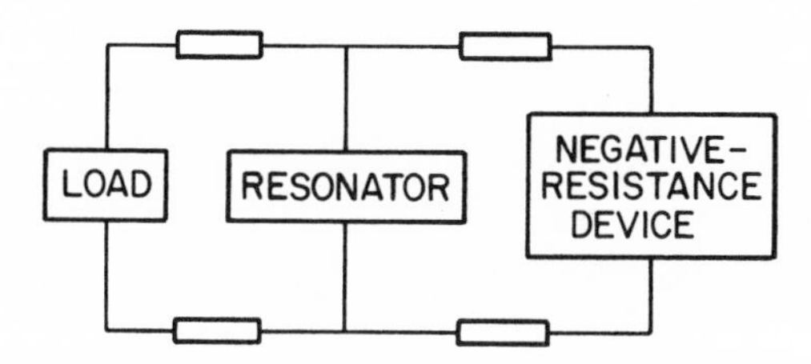

Figure III-9. Equivalent circuit of a negative-resistance microwave two-terminal oscillator.

described in 1963. As TED technology became relatively mature, it was covered in several textbooks [79,80]. We will not describe TED device physics and operating modes.

ii. Impact avalanche transit-time devices (IMPATTs): The IMPATT diode, unlike the TED, is a junction (pn or Schottky barrier) device. The negative resistance in IMPATTs is produced by appropriately combining impact avalanche breakdown and carrier transit-time effects [81]. In principle, since IMPATTs do not rely upon the band structure qualitatively, they can be fabricated from any semiconductor. The technologically important semiconductors are GaAs and Si. The theory of the device was first published by Read in 1958, and the first experimental device was described in 1965. Like the TED, the IMPATT in its many forms is now a relatively mature device and has been described in many textbooks [81] and review articles [82].

a. Device Performance

Figures III-10 and III-11 show the state of the art of cw two-terminal microwave active devices at the end of 1981. These output power and efficiency numbers represent the best laboratory-type devices in a single package. The single package may, however, contain multimesa chips to minimize thermal resistance. The active devices included in the figures are GaAs and Si IMPATTs, and GaAs TEDs. Some results for InP TEDs are also shown. InP TEDs are still in the exploratory stage. The output power and efficiencies for the two-terminal devices are in oscillator circuits. Figures III-10 and III-11 also

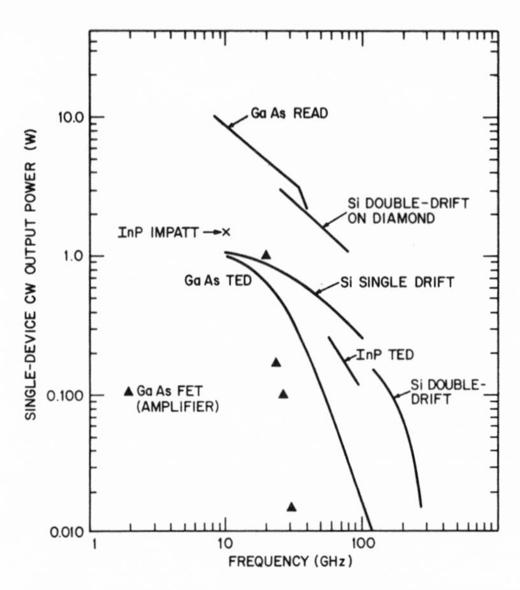

Figure III-10. State of the art (Feb. 1982) of two-terminal cw devices (output power). GaAs FET data shown to indicate inroads made by three-terminal devices.

include some data points for GaAs FETs in amplifier circuits. Note that GaAs FETs are making significant inroads into the frequency range once dominated by two-terminal devices. From figures III-10 and III-11, the following general conclusions can be drawn:

(1) Power levels that can be achieved with TEDs are about an order of magnitude lower than those from IMPATTs. The operating efficiency is also lower. GaAs TEDs have, however, lower AM noise than IMPATTs, especially at modulation frequencies further away from the carrier. This is illustrated in figure III-12 [83]. TEDs are thus more suited to receiver local oscillator (LO) applications, while IMPATTs are better for transmitter applications.

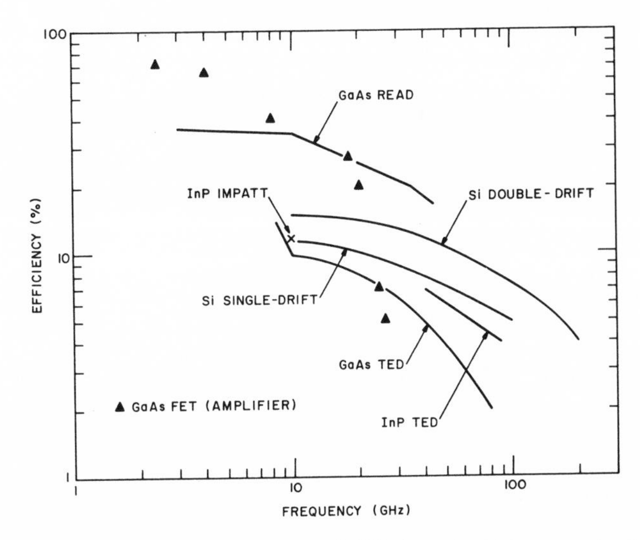

Figure III-11. State of the art (efficiency) of two-terminal cw microwave devices. GaAs FET data also shown to indicate inroads made by three-terminal devices.

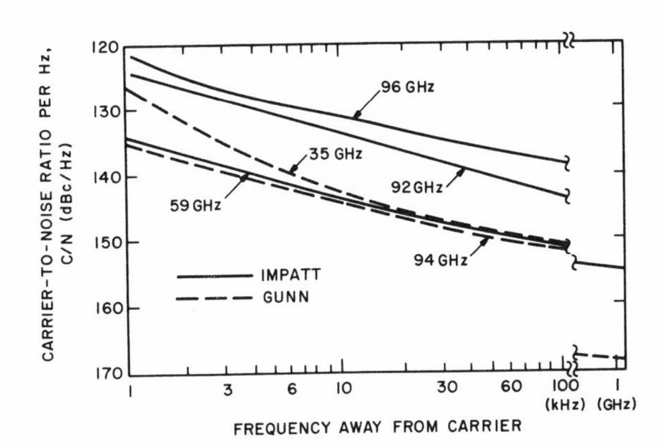

Figure III-12. AM noise characteristics of millimeter-wave IMPATT and Gunn oscillators [83]. Redrawn with permission from Microwave Journal.

(2) GaAs IMPATTs have significantly higher efficiency than Si IMPATTs for frequencies below 35-40 GHz. Si IMPATTs operate at frequencies up to about 250 GHz. In fact, for frequencies greater than 30 GHz, Si IMPATTs and GaAs TEDs represent the only devices that are currently commercially available. From 100 GHz onwards, Si IMPATTs currently reign supreme.

(3) For frequencies over 30 GHz, the performance of developmental InP TEDs lies between that of GaAs TEDs and that of Si IMPATTs.

(4) GaAs FETs are now entering the picture in the 18- to 26-GHz band and will, in general, replace two-terminal devices except perhaps in LO applications.

Figures III-13 and III-14 show low-duty-cycle pulsed characteristics of two-terminal devices. The same general comments also apply for pulsed operation.

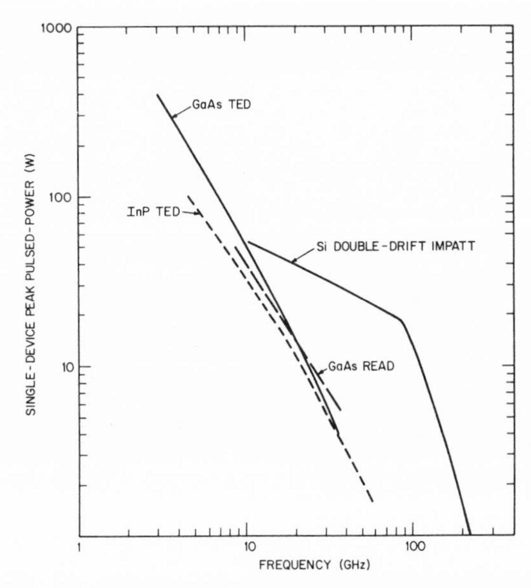

Figure III-13. Low-duty-cycle pulsed operation of single-package, two-terminal microwave devices - peak power.

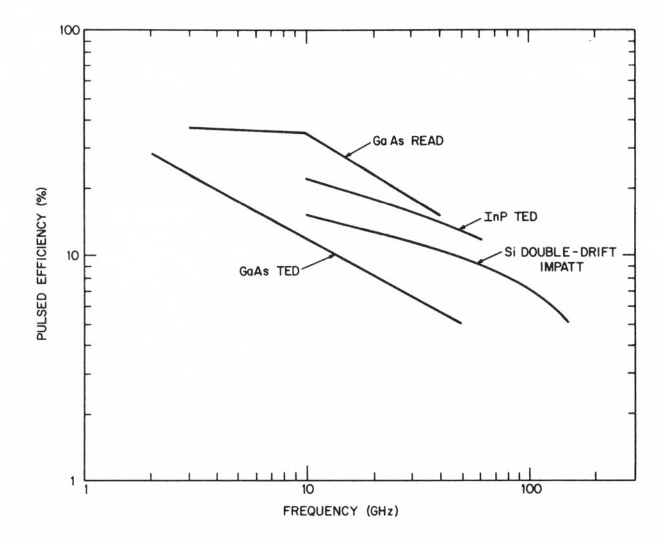

Figure III-14. Low-duty-cycle operation of single-package, two-terminal microwave devices - pulsed efficiency.

b. Amplifiers

Some two-terminal device amplifier considerations are briefly described. We reiterate that when both two- and three-terminal devices are available, use of the latter configuration results in much superior performance.

In general, TEDs exhibit a larger value of negative resistance over a wider bandwidth than do IMPATTs. This implies that a broader bandwidth can be obtained with TED amplifiers. Whereas TEDs operate in a low-constant-voltage, high-current mode, IMPATTs operate at higher voltages and lower (constant) current. As described in Section III-B.1.a (above), IMPATTs have higher output power capability but also have higher noise figures. Table III-13 presents a comparison (not exhaustive - only for illustration) of the performance and specifications of commercially available GaAs TEDs and Si IMPATT amplifiers.

IMPATTs and TEDs can operate as either stabilized amplifiers or injection-locked oscillators (ILO). In general, stabilized amplifiers provide wider bandwidth (5-10%), lower gain, lower output power, and better linearity; ILOs

Table III-13. COMMERCIAL Si-IMPATT AND GaAs TED AMPLIFIERS* [84]

Amplifier	26-40 GHz	40-60 GHz	50-75 GHz	75-110 GHz
Si IMPATT	200 mW	100 mW	60 mW	20 mW
GaAs TED	100 mW	40 mW	20 mW	--

Other Typical Specifications	Si-IMPATT Amplifiers	GaAs TED Amplifiers
Gain (dB min)	16	16
Bandwidth (GHz min)	0.5	1.0
Efficiency (% typical)	1.0	0.5
Harmonic Content (dB typical)	30	15
Amplitude-Modulation to Phase-Modulation Conversion (°/dB typical)	5.0	5.0
Spurious Responses (dBc)	60	60
Noise Figure (dB typical)	40	22
Operating Voltage (V max)	40	7.0
Operating Current (A max)	0.5	3.0

*For illustrative purposes only; not exhaustive list.

are more suited for narrowband, high-gain, higher-output-power applications. Examples of systems suited for ILO are radar and FM communication systems. At millimeter waves, where IMPATTs currently find most application, ILOs are easier to implement than stabilized amplifiers [83]. This is particularly true when IMPATTs with more complex doping profiles (double-drift, Read) are used. ILOs follow the classical rule of $\sqrt{G} \times BW$ (bandwidth) = constant; a gain-bandwidth product of 3-6 GHz is typical at millimeter waves [84].

Table III-14 provides examples of IMPATT amplifiers reported in the literature [78]. Note that they all are narrow-bandwidth amplifiers for FM microwave radio applications. The noise figures are in the 26- to 50-dB range, and overall efficiencies are less than 10%. To provide a contrast, we have also included some published results on medium-power GaAs FET amplifiers. Note that output powers as high as 25 W with over 18% efficiency and AM/PM conversion of about 1°/dB at 5 GHz have been achieved [85]. The performance of the GaAs FET amplifier developed by RCA in the 3.7- to 4.2-GHz band is even more impressive [86].

Table III-14. IMPATT AMPLIFIER PERFORMANCE*

	Frequency (GHz)	Output Power (W)	Gain (dB)	Noise Figure (dB)	Overall Efficiency (%)	Stages	Diode Type
IMPATT Amplifiers [78]	6.0	1.0	20	48	4	1	Si (Flat)
	6.0	1.6	22	43	7	1	GaAs (Flat)
	6.0	10	30	30	7	1	GaAs (Flat)
						2,3	2-GaAs (Flat)
	11.0	3.5	25	36	6	1,2,3	GaAs (Flat)
	6.0	10	40	26	6	1,2,3	Gunn
						4,5,6	GaAs (Flat)
						7	Modified-Read
						8	2-Modified-Read
	13.5	5	57	--	10	1,2,3	GaAs (Flat)
						4,5,6,7	Modified-Read
GaAs FET Amplifiers [85]	4.9-5.2	25	29	--	18		
[86]	4.0-4.2	8.5	53	--	33		

*GaAs FET amplifier data shown to provide comparison.

At frequencies higher than 35 GHz, two-terminal Si devices still dominate. While two-terminal InP devices and three-terminal GaAs devices have begun to enter the 26- to 40-GHz band, Si IMPATTs represent the established technology in this band and at higher frequencies. A lot of work on devices, amplifiers, and passive power combiners is being carried out in the 35-, 40-, 60-, 94-, and 140-GHz regions. This work was reviewed by Kuno in 1981 [83].

2. Three-Terminal Microwave Devices - Microwave Transistors

Discrete Si n-p-n bipolar transistors began making significant contributions to microwave systems in the mid-to-late 1960s. Cost-effective transistors with proven reliability are commercially available for low-noise amplifiers, low-distortion linear amplifiers, high-power Class C amplifiers, and oscillators. This holds true especially for frequencies below 3 GHz. An excellent, albeit slightly partisan, review of the status and prospect of microwave Si bipolar transistors has been given by Snapp [87]. At frequencies above 5 GHz, the inherent performance limitations of Si bipolars have prevented their viability. Even in the 2- to 5-GHz range, the Si bipolar transistor has been successfully challenged by GaAs FETs, particularly in medium-power (∿10 W) linear amplifier applications [88]. We will not discuss Si bipolar transistors any further; their performance will, however, be used as a baseline for the discussion of GaAs FETs.

a. GaAs FETs

The discrete GaAs FET technology has matured rapidly over the last decade. An excellent tutorial review of GaAs FETs was given by Liechti in 1976 [89]. As a result of intensive effort, GaAs FET technology has progressed from the early demonstration phase to second- and third-generation processes and is finding many system applications. Nippon Electric 244 series low-noise FETs have been space-qualified and deployed in space (ANIK B). RCA has developed and space-qualified an 8.5-W, 55-dB gain, 3.7- to 4.2-GHz amplifier using GaAs FETs for the commercial advanced SATCOM satellite that will be launched late in 1982 [86].

The popularity of the GaAs FET is a consequence of its superior frequency response, low noise figure, high linearity, good power-added efficiency, and versatility. Many laboratories are currently developing low-noise GaAs FETs with 0.25-μm gate lengths that operate up to 30 GHz with noise figures as low as 4 dB, with 5 dB of associated gain [90]. RCA is developing power GaAs FETs that have demonstrated 20 dBm of output power with a 3-dB gain at 26 GHz. Work at RCA has shown GaAs FETs to be capable of operating with power-added efficiency as high as 72% at 2.45 GHz at a cw power level of 1.2 W.

i. GaAs FET Design Considerations: The GaAs FET is a majority-carrier device that operates by controlling the current flow through a conducting channel by means of a Schottky-barrier control gate. The basic device configuration and its equivalent circuit elements are shown in figure III-15(a). The operation of the device at high microwave frequencies requires the gate length to be less than 1 μm. The fundamental material properties of GaAs, such as high low-field mobility, high peak electron velocity, and the existence of an insulating ($\sim 10^6$-10^8 Ω·cm) GaAs substrate, are of major importance for microwave operation.

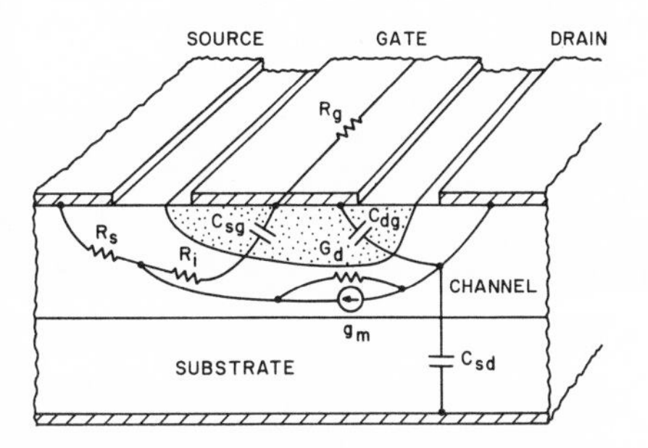

(a) Equivalent circuit of a Schottky-barrier FET superimposed on the cross section of the transistor.

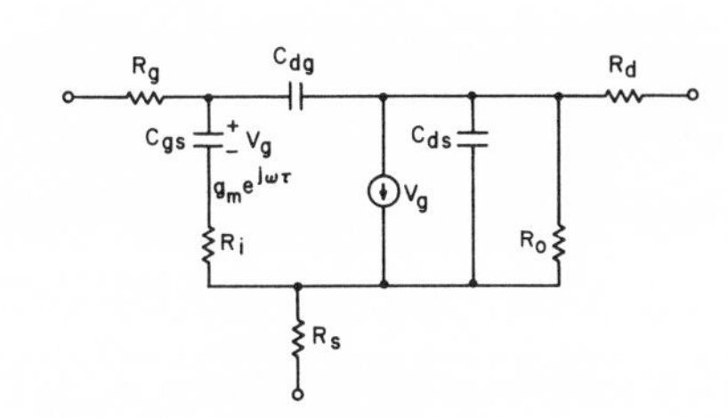

(b) FET equivalent circuit.

Figure III-15. GaAs FET configurations.

A comprehensive analysis of GaAs FETs based on an analytic one-dimensional model was carried out by Pucel et al. [91]. While a simple form for the GaAs velocity-field characteristic is used, this model (known as the PHS model) predicts dc and small-signal parameters accurately enough for first-order device design. As the device gate lengths decrease, there is a possibility that the electrons do not reach equilibrium with the lattice. As a consequence, in short(submicrometer)-channel devices the electron velocity may actually be higher than the peak average velocity, leading to smaller transit-time and thus higher cut-off frequency [92]. The implications of these nonequilibrium effects are discussed in Section III.B (Exploratory Development). Figure III-15(b) shows the equivalent circuit of the GaAs FET.

A number of analyses of the noise behavior of GaAs FETs have been performed. The initial work was done by Van der Ziel [93], who predicted the noise behavior of the Shockley transistor. Later, Baechtold included the effects of velocity saturation by assuming that the average electron temperature in the channel is somewhat hotter than the lattice temperature [94]. The noise contribution of the velocity-saturated region of the device was, however, not calculated. This was subsequently done by Statz, Haus, and Pucel, who were the first to obtain good agreement with experiment [95]. These authors showed that the noise contribution of the high field region was due to drifting dipole layers in the saturated drift velocity zone. Such diffusion noise was shown to be the dominant contribution under certain bias conditions. Use of this approach made it possible for the first time to correctly predict the dependence of noise figure on drain current.

No noise analyses that take nonequilibrium effects into account have been published. This problem is very complex, but rough predictions of the noise figure of short-gate devices can be made by extrapolating available analyses to short gate lengths and then inserting the gain-bandwidth product obtained from velocity overshoot (nonequilibrium) calculations into the noise figure expression.

In 1979, an alternative formulation of FET noise figure analysis was published by Fukui [96]. This work, based on empirical curve fitting to device characteristics, provides reasonable design guidelines for low-noise device optimization.

The major conclusions of the PHS calculations of device equivalent circuit parameters are summarized below:

(1) Under normal operating biases and channel thicknesses, most of the channel is in velocity saturation.

(2) A substantial fraction of the gate capacitance is due to parasitics. This fraction increases as gate length decreases.

(3) The transconductance depends directly on the saturated drift velocity of electrons. (Any increase in electron velocity due to nonequilibrium effects increases the transconductance.)

(4) Parameters, including transconductance (g_m), gate-source capacitance (C_{gs}), and gate resistance (R_i), are essentially fixed once gate length, materi[al] parameters, and source-to-gate spacing are chosen, and are not influenced strongly by geometrical layout or fabrication technology.

(5) Parasitic resistances (R_s) and (R_d), the gate metallization resistance (R_g), the drain resistance (R_o), and the drain gate capacitance (C_{dg}) are dependent on device design, construction, and channel material quality. Efforts to improve performance are therefore concentrated on controlling these parameters.

(6) At frequencies higher than X-band, extrinsic parasitics such as lead inductance and package capacitance become increasingly important.

The goal of this discussion is only to indicate factors affecting FET design. Details can be found in the literature [91]. Fukui's noise figure analysis is discussed briefly in the following section on low-noise FETs.

ii. Low-Noise GaAs FETs - 300 K Operation: The potential of GaAs FETs for low-noise applications is illustrated in figure III-16, where the best published noise figure data (late 1981) are shown as a function of frequency. Noise figure data of Si bipolar transistors are also shown to illustrate the significantly superior performance of GaAs FETs. Figure III-16 includes the theoretical noise figure computed by Fukui in 1979 [96] for a 0.25-μm gate-length, 65-μm gate-width GaAs FET optimized for low noise. For frequencies up to 18 GHz, the experimental data are close to the computed curve; at higher frequencies, where optimal noise matching is more difficult to achieve and circuit losses are higher, the experimental noise figures are also higher.

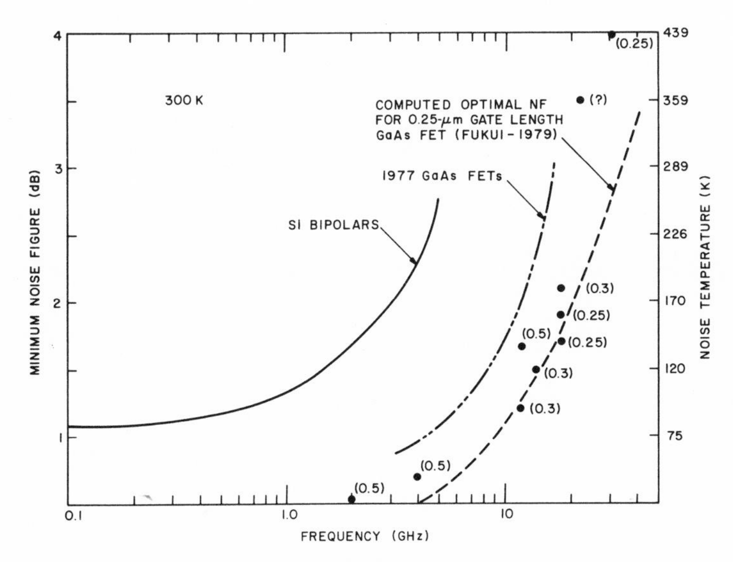

Figure III-16. Minimum noise figure of GaAs FETs (late 1981 data). Number in parentheses is gate length. Best 1977 data shown to indicate progress. Theoretical curve for 0.25-μm gate-length GaAs FET also shown.

For solid-state devices at microwave frequencies, the gain is relatively low, and a better indicator of noise performance is the noise measure M [97]. The noise measure, first introduced by Haus and Adler [97], is defined as

$$M = 10 \log_{10} (F - 1)/[(1 - 1/G)]$$

where F is the noise figure (ratio) and G is the associated gain (ratio). Figure III-17 repeats the GaAs FET data of figure III-16 but noise measure is displayed as a function of frequency. To obtain the lowest possible system noise figure, the first-stage device should be selected for the lowest noise measure. Table III-15 lists the noise figure, associated gain, and noise measure of some commercially available GaAs FETs.

To put these data into perspective and look at development trends, we briefly consider the low-noise FET optimization analysis of Fukui [96]. The noise performance of linear two-ports is characterized by four noise parameters: the minimum noise figure, equivalent noise resistance, optimum source resistance, and optimum source reactance. An active two-port device usually consists of the intrinsic part and surrounding parasitics. Each of the noise parameters of this device can be expressed in terms of the intrinsic device parameters and extrinsic parasitic parameters. The parasitic parameters are then classified into two categories, resistive and reactive. In microwave transistors, the minimum noise figure is relatively insensitive to the reactive parasitics, while the other noise parameters are strongly affected [96]. As the operating frequencies increase, these three noise parameters become increasingly influenced by the reactive parasitics, such as lead inductances and package capacitances. In contrast, the frequency characteristic of the minimum noise figure is reasonably well described regardless of the reactive parasitics.

Fukui shows that the minimum noise figure F_o can be written in terms of device geometrical and material parameters [96]. The equation for F_o is

$$F_o = 1 + kf \left(\frac{NL^5}{a}\right)^{1/6} \left[\frac{17\, z^2}{hL_g} + 1.32\, z^2 \left(\frac{f}{hL_g}\right)^{1/2} + \frac{2.1}{a_1^{1/2} N_1^{2/3}} + \frac{1.1\, L_2}{a_2 N_2^{0.82}} + \frac{1.1\, L_3}{a_3 N_3^{0.82}}\right]^{1/2} \qquad (3)$$

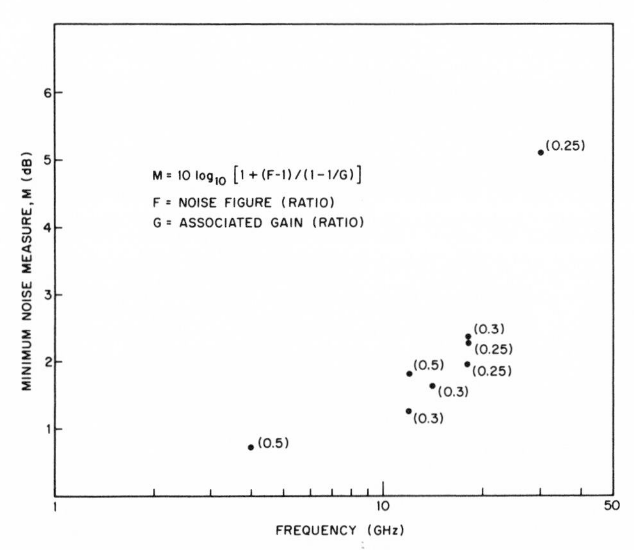

Figure III-17. The data of figure III-16 converted to noise measure.

The parameters for equation (3) (see fig. III-18) are defined as follows:

N = carrier concentration in channel ($x10^{16}$ cm^{-3})

N_1, N_2, and N_3 = carrier concentration in regions 1, 2, and 3, respectively ($x10^{16}$ cm^{-3})

L = channel length (μm)

L_1, L_2, and L_3 are lengths (μm) of regions 1, 2, and 3, respectively

L_g = gate metal length (μm)

f = operating frequency (GHz)

h = thickness of gate metallization (μm)

a = channel thickness under gate (μm)

a_1, a_2, a_3 are thicknesses (μm) in regions 1, 2, and 3, respectively

k $\simeq$ 0.04, an experimentally determined constant

z = unit (gate-stripe) gate width (mm)

Table III-15. COMMERCIAL LOW-NOISE GaAs FETs (Dec. 1981) - TYPICAL PERFORMANCE*

Manufacturer	Type No.	Noise Figure, F (dB)				Associated Gain, G (dB)				Noise Measure, M** (dB)			
		4 GHz	8 GHz	12 GHz	18 GHz	4 GHz	8 GHz	12 GHz	18 GHz	4 GHz	8 GHz	12 GHz	18 GHz
Alpha	ALF3000 (C)†	0.9	1.5	2.1		13.0	11.0	8.5				2.36	
	ALF3003 (P)†	0.9	1.5	2.1		13.0	10.5	8.0				2.40	
Avantek	AT8040 (P)			2.4				9.0				2.67	
	AT8041 (C)			2.0	2.8			10.0	7.0			2.17 (5) ††	3.29 (2)
	AT8060 (P)			2.8				8.0				3.11	
	AT8061 (C)			2.5	3.3			9.0	5.0			2.76	4.26
Dexcel	DXL2503A (C)		1.9	2.8	4.0		10	8	5.5			3.17	4.92
Mitsubishi	MGF1400 (P)	2.0	3.2	4.5		9	7	6				5.35	
	MGF1402 (P)	1.1	2.0	3.0		13	10	8				3.39	
	MGF1403 (P)	0.8	1.3	1.8	2.8	14	12	10.5	7			1.94 (2)	3.29 (2)
	MGF1412 (P)	0.8				13							
NEC	NE13700 (C)	0.8	1.2	1.9	2.5	14	11	9.5	7.5			2.09 (3)	2.89 (1)
	NE13783 (P)	0.8	1.2	1.9		14	11	9				2.12 (4)	
	NE13783S† (P)	0.8	1.2	1.6		14	11	9.5				1.76 (1)	
	NE38800 (C)	1.2	2.0	2.8	4.0	13.5	11.5	8.5	6.0			3.13	4.80
	NE38883 (P)	1.2	2.3	3.2		13.0	11.0	7.5				3.66	
	NE70000 (C)	1.0	1.5	2.3		13	11	9				2.55	
	NE70083 (P)	1.0	1.5	2.3		13	11	9				2.55	
Plessey	GAT 6/P100 (C)	1.0	1.5	2.2	3.5	15	11.5	9.5	7.0			2.41	4.06
	GAT 6/P104 (P)	1.0	2.0	2.5		14	9.5	8.5				2.80	

*Compiled by R. E. Askew (RCA).

**$M = 10 \log_{10} \left[1 + \frac{F - 1}{1 - 1/G}\right]$

†(C), chip; (P), packaged; S, selected by manufacturer for an optimum noise figure at 12 GHz.

††(#), rank.

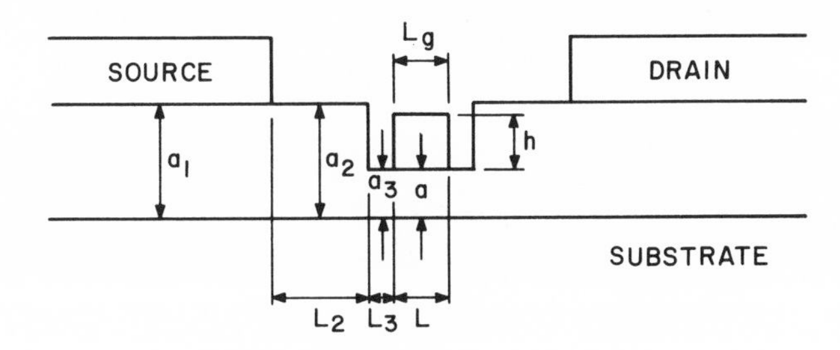

Figure III-18. FET cross section.

This expression assumes Al metallization for the gate and includes skin effect.

It is instructive to write F_o in terms of the equivalent circuit parameters [96],

$$F_o = 1 + 0.27\ Lf\ \sqrt{g_m(R_g + R_s)} \quad \text{and}$$

$$g_m \simeq 0.023\ \alpha z\ (N/aL)^{1/3}\ S \tag{4}$$

where g_m is transconductance, α is the number of gate stripes of width z in $\alpha z = Z$, the total gate width (mm). From equations (3) and (4) for F_o, the following statements can be made:

(a) To decrease the minimum noise figure F_o, it is necessary to reduce gate length L_g, source parasitic resistance R_s, and gate metal resistance, R_g.

(b) Source (and drain) parasitic resistances can be lowered by increasing doping under contacts ($\sim 10^{18}$ cm^{-3} n^+ regions), narrowing the source-gate spacing, and increasing the GaAs thickness under source and drain contacts (recessed-gate geometry).

(c) Gate parasitic resistance can be decreased by increasing metal thickness h, decreasing unit gate width z, and increasing gate length L_g. Increasing the gate length is obviously counterproductive.

Note that decreasing the gate length by itself is not useful - it is also necessary to increase the gate metal thickness and/or reduce the unit gate width. It is technologically difficult, for example, to reduce L_g to 0.25 μm and increase h to more than 0.5-1 μm.

Figure III-19, taken from Fukui's paper [96], illustrates some of these points. Note that going from a 0.9- to a 0.5-μm gate length, keeping z at 0.25 mm (curves b and c) does not decrease F_o significantly. Curves c and d illustrate that keeping L_g at 0.5 μm and decreasing z from 0.25 to 0.1 mm improves F_o significantly. Figure III-20 shows a calculation by Hughes Research Laboratory [98] of the 12-GHz noise figure of a 0.25-μm gate-length GaAs FET as the gate metal thickness h is varied. Note the dramatic improvement in F_o as h is increased.

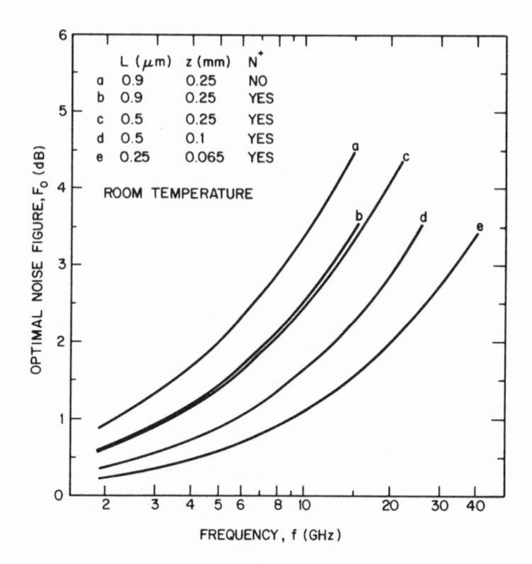

Figure III-19. Noise figure of GaAs FETs as a function of frequency computed for different device parameters by Fukui [96]. © 1979 IEEE.

An important point is that the minimum noise figure F_o depends only on unit gate stripe width; the transconductance and hence gain, however, depend on total gate width. Since high gain and minimum noise figure are both important (i.e., to minimize the noise measure M), a multistripe gate geometry is usually preferred.

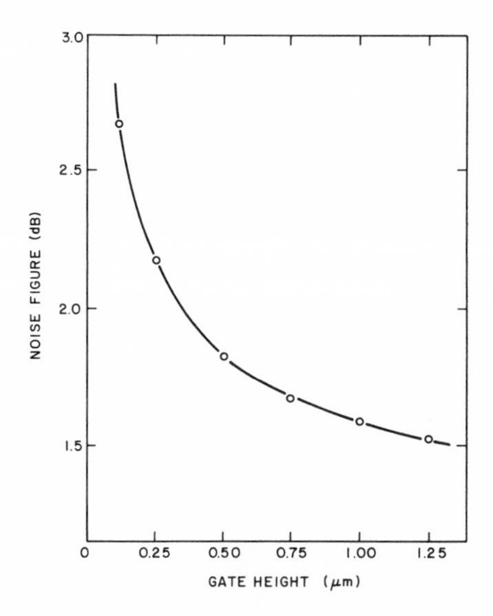

Figure III-20. Noise figure vs gate metallization thickness for 0.25-μm FET.

Fukui also points out that a quick estimate of the minimum noise figure of Si bipolar transistors at 300 K is given by the expression

$$F_o|_{min} = 1 + bf^2 \left[1 + \left(1 + \frac{2}{bf^2} \right)^{1/2} \right] \quad (5)$$

where $b = 40\ I_c r_b / f_T^2$. I_c is the collector bias current (A) and r_b is the parasitic base resistance (Ω). For the GaAs FET,

$$F_o|_{min} = 1 + mf$$

where

$$m = \frac{2.5}{f_T} \left[g_m (R_g + R_s) \right]^{1/2} \quad (6)$$

Thus, the noise figure of the Si bipolar transistor increases quadratically with frequency, while that of GaAs FETs increases linearly.

The following trends for conventional planar GaAs low-noise FETs may be anticipated:

(1) Work on the optimization of device geometry will continue. Major attention will be given to minimizing parasitic resistances, particularly the gate metal resistance. Use of thicker metallization and T-geometries are some possible alternatives. This trend will be limited by technological constraints.

(2) We believe that 0.25-μm gate length is close to the point of diminishing returns as far as gate-length minimization is concerned. For frequencies below Ku-band, optimized 0.5-μm gate-length devices may be adequate.

(3) The use of selective ion implantation to generate n^+ source and drain regions will become more prevalent.

(4) Evolutionary improvement in material quality and manufacturing learning-curve effects will improve the noise figure of commercial devices, bringing them closer to "best" laboratory results.

(5) More low-noise "gain blocks" consisting of monolithic two- or three-stage amplifiers will probably come on the market. This procedure will minimize the packaging parasitics in the critical front end of receivers.

(6) GaAs FETs operating in the EHF band will become commercially available in the 1984 time frame.

iii. Cryogenic GaAs FET Amplifiers: There are many applications, particularly those requiring large aperture antenna systems such as space communications and radio astronomy, where no current device operating at 300 K has sufficiently low noise. Cooled GaAs FET amplifiers can find use in many of these applications. An excellent discussion of this topic was presented by Weinreb in 1980 [99].

Figure III-21, taken from Weinreb [99], shows the 1980 state of the art for microwave low-noise amplifiers at both 300 K and cryogenic temperatures. At 20 K, the GaAs FET amplifier performance is almost comparable to that of parametric amplifiers (paramps). In addition, GaAs FETs have two inherent advantages: (1) The GaAs FET is much less critical to circuit impedance than a negative resistance amplifier such as a paramp. (2) The GaAs FET amplifier is powered by dc in contrast to paramps that need a power oscillator and tuned circuits at several times the operating frequency.

Figure III-22, again taken from Weinreb [99], shows the noise temperature of several commercial GaAs FETs (1979-1980 technology) at 4.5 GHz as a function of physical temperature. Weinreb also points out that cooling below 15 K does

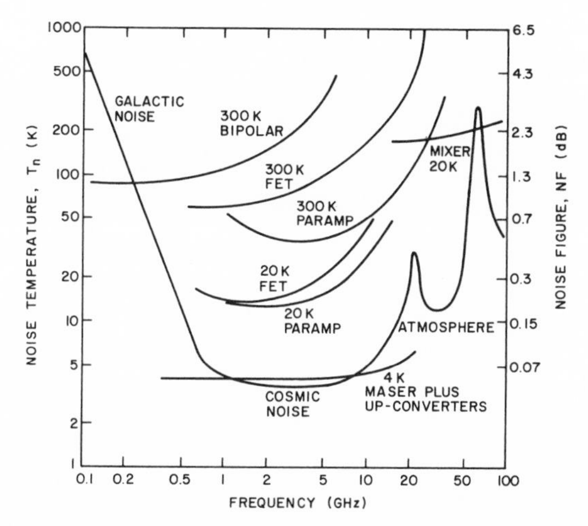

Figure III-21. Cryogenic low-noise amplifier (~1979-1980) [99]. © 1980 IEEE.

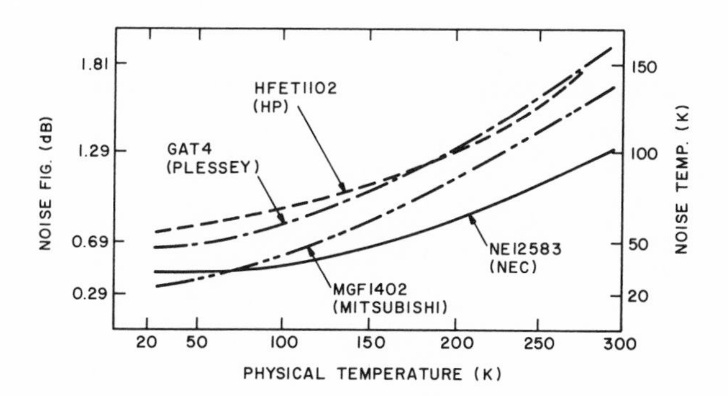

Figure III-22. Noise figure of commercial low-noise GaAs FETs (1979-1980) as a function of temperature [99]. © 1980 IEEE.

not decrease noise temperature significantly since the thermal conductivity of GaAs decreases due to boundary scattering of phonons, and self-heating effects will dominate [99].

b. GaAs Power FETs

The microwave performance of GaAs power FETs has improved rapidly over the last few years as a result of intensive effort at several laboratories all over the world. Multichip devices that generate 20-30 W at C-band to single-chip structures generating 100 mW at 26 GHz have been demonstrated. As indicated earlier, RCA has developed and space-qualified 8.5-W, 55-dB gain, 33% overall efficiency GaAs FET amplifiers for SATCOM application in the 3.7- to 4.2-GHz band [86]. Researchers at Bell Laboratories are developing GaAs power amplifiers for radio relay applications. A 25-W GaAs FET amplifier in the 4.9- to 5.2-GHz band for a microwave landing system has been described in the literature [85]. The GaAs FET power amplifier is a very strong contender to replace medium-power (10-20 W) TWTAs in many communications applications. Table III-16 qualitatively describes the advantages of GaAs FET amplifiers compared to those of TWTAs.

Table III-16. ADVANTAGES OF FET AMPLIFIERS OVER TWTAs

- Lower Intermodulation Distortion
- More-Ideal Saturation Characteristics
- Smaller Variation in Phase Shift as a Function of Drive
- Lower AM/PM Conversion
- No Fine-Grain Gain Variations
- Longer Life

The replacement of TWTAs by GaAs FETs not only offers the time-honored advantages of solid-state devices such as smaller size and potential reliability, but can also result in significant performance enhancement. For example, in linear amplifiers for communications such as in multicarrier service, or for low-phase distortion amplifiers in FM-FDM service, the GaAs FET amplifier is superior to TWTAs [86]. For these applications, the TWTAs are generally operated in a back-off mode, leading to a large reduction in the TWTA dc-to-rf conversion efficiency. These tradeoffs are discussed by Dornan et al. of RCA in reference 86.

To put the state of the art of GaAs FETs in perspective, figure III-23 shows the typical cw and pulsed output power capability of commercial Si bipolar transistors as a function of frequency. A few data points for laboratory GaAs FETs are also shown. Note that the Si bipolar data are for Class C operation, while the GaAs FET data are for Class A (or AB) operation. Figure III-24 shows the output power-frequency data for both laboratory and commercial GaAs FETs. Note that commercial devices with about 0.75 W output power at 15 GHz are available (Feb. 1982 data), and experimental devices operating up to 30 GHz have been reported.

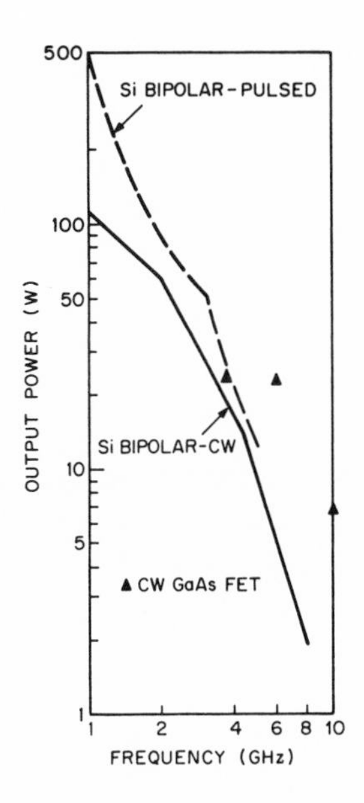

Figure III-23. Si-bipolar-transistor (classic amplifier) state of the art. GaAs FET (Class A or AB) points shown for reference.

i. Power GaAs FET Design Considerations: GaAs power FETs have many similarities with the low-noise devices but also have significant differences. Both similarities and differences are listed below.

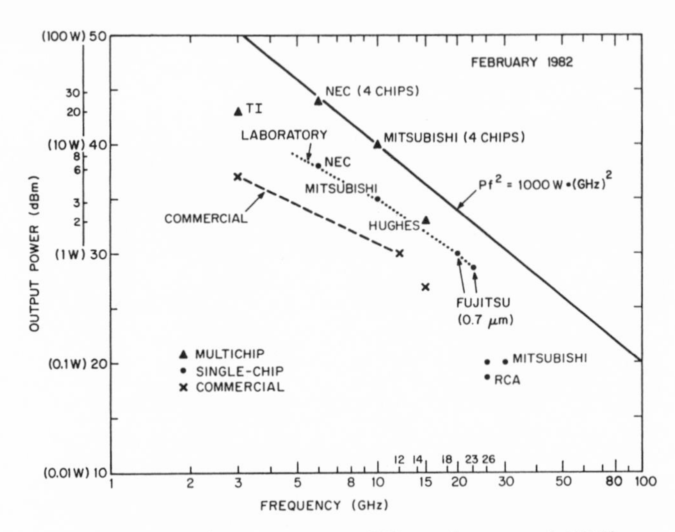

Figure III-24. GaAs power-FET performance (~1982).

(1) The GaAs power FET operates at a much higher source-drain bias voltage than its low-noise counterpart. Operation at higher bias voltage is necessary to allow a sufficiently large voltage swing to generate the desired output power. The cut-off frequency of a GaAs FET generally decreases as the source-drain bias (V_{DS}) is increased [100]. This implies that at high microwave frequencies, the power FET is, in general, more difficult to realize than its low-noise counterpart. Furthermore, this requirement implies that the source-drain burnout voltage must be maximized. Source-drain burnout maximization is therefore a major research topic.

(2) The gate width of the power FET has to be made larger than that of a low-noise device. The gate width must be increased to provide sufficient current capability to obtain the required output power. To keep the gate metallization resistance small, a multistripe (interdigitated) geometry is employed to increase the gate width. Increasing the gate width also reduces the device impedance, making impedance matching difficult. This problem gets worse as the frequency increases. The approach usually taken is to determine the optimum gate width at a given frequency (or bandwidth) from impedance

considerations and to configure the device with "cells" of the optimum gate width. These cells can then be prematched to 50 Ω or any convenient impedance and then interconnected. This cell interconnection can be accomplished on the chip carrier (package) or in a monolithic format.

(3) As in the case of all power devices, the thermal impedance of the device must be minimized by proper heat-sinking. This requirement has led to several power FET designs (fig. III-25) that will be discussed presently.

(4) The device and mounting parasitics must be minimized. This requirement becomes more difficult to meet as the operating frequency increases. The source inductance and gate metallization resistance are two important parasitics that must be minimized.

A review of GaAs power FET design and performance was presented by DiLorenzo and Wisseman [101] in 1979. We briefly present part of their simplified analysis to illustrate some key points. The four points discussed above will then be amplified to illustrate current research trends.

Figure III-26 shows the simplified static I-V characteristics of a GaAs power FET. Curve C_1 shows the maximum device current I_f when the Schottky barrier gate is forward-biased and the channel is completely open. Curve C_2 is the load line whose slope is the reciprocal of the load resistance. The point corresponding to V_{SD}, I_{DS}, V_G is the quiescent operating point. The other important parameters are the knee voltage V_k and the limiting source-drain voltage V_{SD}^L, and V_p is the terminal pinch-off voltage (i.e., the channel is pinched off when $V_G = -V_p$). This limiting voltage is set by the gate-drain avalanche breakdown voltage V_{GD}, as in the expression $V_{SD}^L = V_{GD} - V_p$.

Beyond this point, excess current occurs in the drain circuit due to avalanche breakdown that cannot be modulated by the gate and therefore does not contribute to device output power. The maximum output power can then be calculated to be

$$P_m = \frac{I_f}{8} (V_{GD} - V_p - V_k) \qquad (7)$$

DiLorenzo and Wisseman claim that P_m is about 1 W/mm of gate width. In general, our experience shows that P_m is about 0.5-0.8 W/mm for most available devices. For frequencies about Ku-band, P_m is usually 0.5 W/mm and decreases as frequency increases. For optimum output power, I_f and V_{GD} must be increased and V_k decreased. These parameters are interrelated: an increase in

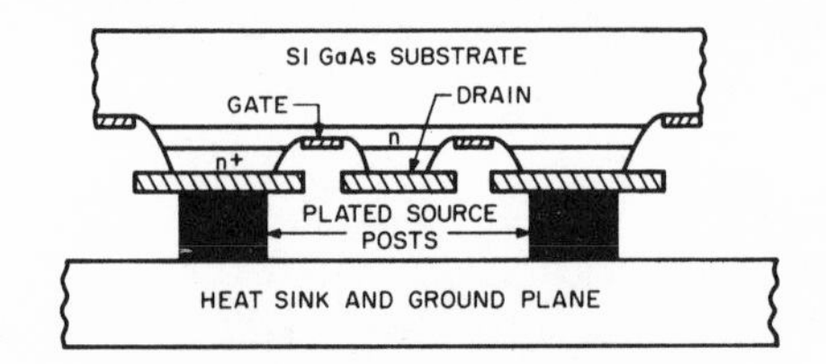

(a) Flip-chip-mounted GaAs FET.

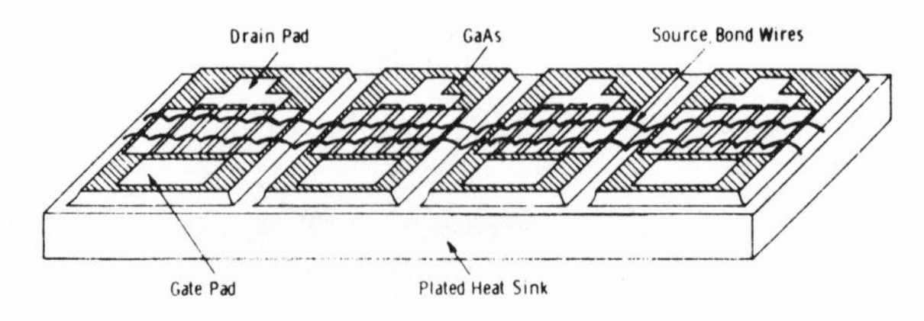

(b) Plated-heat-sink FET.

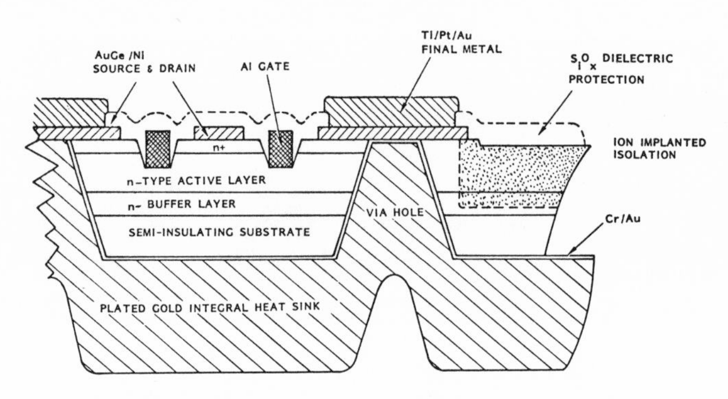

(c) Via-hole source-grounded power GaAs FET.

Figure III-25. Representative FET geometries [101]. © 1979 IEEE.

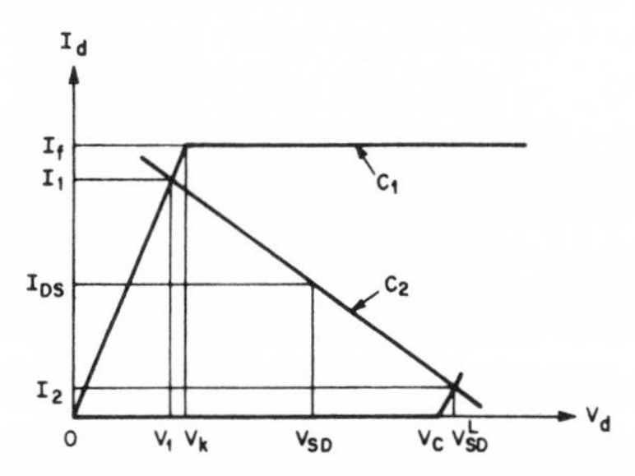

Figure III-26. Schematic representation of the I-V characteristic of a power GaAs FET showing the important variables for power [101]. © 1979 IEEE.

I_f by an increase in doping decreases the breakdown voltage V_{GD}. At constant doping, an increase in I_f increases V_p, and so on. Thus tradeoffs have to be made, and these are usually determined empirically.

Gate-Drain Avalanche Breakdown. As indicated earlier, the onset of gate-drain avalanche imposes a constraint on the rf voltage swing and, hence, on the output power capability of GaAs power FETs. Wemple et al. [102] have shown empirically that the recognition of the surface depletion phenomenon in GaAs and minimization of the undepleted charge per unit area between the gate and drain can result in avalanche voltage factors of 2-3 above the bulk value. This is a result of the two-dimensional character of the channel electric field. Computer simulation by Curtice [103] confirms these conclusions. Recessing the gate by the surface depletion thickness and keeping the channel saturation current (prior to gate recessing) below 400 mA/mm of gate width is one appropriate strategy for minimizing breakdown [102]. Technological considerations such as well defined, uniform gate metallization without sharp edges, etc., are of equal importance.

Gate-Width Considerations. The strategies currently being used in various laboratories to increase gate width have been reviewed by DiLorenzo and Wisseman [101]. These strategies comprise various methods for paralleling a number of gate stripes.

The planarity of the GaAs FET structure (i.e., source, drain, and gate contacts lying almost on a plane) requires that some means of isolation must be provided to prevent two of the three electrodes from short-circuiting. Fujitsu [104] and NEC [105] have used a gate-over-source crossover with dielectric

(SiO_2) isolation. This strategy makes efficient use of GaAs real estate at the expense of processing complexity. Fujitsu has developed a 26-mm-wide device.

Paralleling has been achieved by interconnection with wire bonding (TI), air-isolated metal bridging (RCA), flip-chip-bonding plated-up source pads (RCA, MSC, and Mitsubishi), and flip-chip-bonding metal-bridged sources (RCA, Mitsubishi). Flip-chip mounting also provides excellent heat-sinking and minimal parasitic source inductance. The flip-chip approach was pioneered by RCA [106].

However, increasing the gate width by paralleling results in reduced device impedance and makes circuit matching difficult. The impedance matching problem becomes more difficult at higher frequencies or when large bandwidths are required. As mentioned above [see under (2) in Section III.B.2.i], for such applications the preferred strategy is to prematch cells of optimum width and then interconnect cells in the package or in a monolithic format. The optimal cell width is a function of frequency, desired bandwidth, and to a certain extent the specific device geometry. For in-package matching, wire bonds can be used as inductors, and package standoffs as capacitors. Figure III-27 is a photograph of an on-carrier (package) matched amplifier. This configuration was developed by R. L. Camisa of RCA [107].

Thermal Management. The thermal design of a power device is of obvious importance. Thermal analysis depends on the specific device configuration and requires a three-dimensional heat-flow calculation. The thermal design of power FETs has been described by Wemple and Huang [108]. The following general statements can be made:

(1) For power FETs that are mounted upright on the heat sink, heat must be removed through the GaAs substrates. With the relatively low thermal conductivity of GaAs, the substrate must be thinned. The minimum thickness is determined by technological considerations such as ease of thinning, breakage due to fragility of GaAs, and so on. It is difficult to thin the substrate to less than about 50 μm. After thinning, a metal heat sink can be plated on.

For multigate FETs (almost all power FETs), the gate-to-gate separation (i.e., the separation between adjacent heat sources) is another key design parameter. To prevent the thermal interaction between adjacent sources, it is desirable to increase their separation. Clearly, the thicker the substrate the more important the adjacent gate separation. Other considerations such as wavelength constrain this separation at higher microwave frequencies.

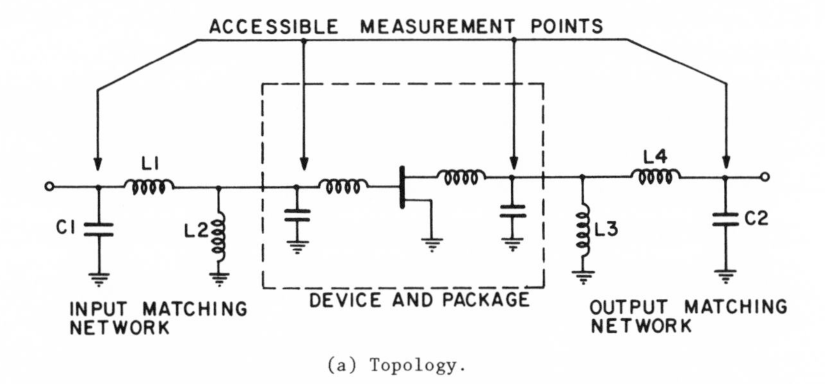

(a) Topology.

Output
C2
L4
L3
Flip-Chip FET
1.5 mm
High Value Capacitors (RF Ground)
Copper Carrier
C1
L1
L2
Input

(b) Realization [107].

Figure III-27. Lumped-element amplifier [107].

(2) For flip-chip-mounted FETs, the heat is removed through the source posts [108]. The spreading resistances between GaAs and the source posts and those between the post and heat sink are important [see fig. III-25(a)]. Again,

gate-to-gate separation, source pad size, and integrity of the flip-chip bond are important.

It is significant that in most power FET designs, the temperature rise is about 50-70°C above ambient. This is considerably lower than that of solid-state devices like the IMPATT (with a temperature rise of ∿200°C).

Parasitic Minimization. The minimization of parasitics is very important for GaAs FETs. We briefly discussed the effect of intrinsic parasitics on the frequency response of low-noise FETs; these considerations also apply to power devices. In addition, extrinsic parasitics, such as the source inductance (L_s) and mounting capacitances due to package standoffs, become important, particularly at higher frequencies. The device total gate width and the desired bandwidth also influence the amount of extrinsic parasitics that can be tolerated. Some strategies that have evolved and are being pursued are sheet grounding of the source pads, via-hole grounding of sources [109], and flip-chip mounting of the sources to a good rf ground [106]. Each method has both advantages and disadvantages, and it is difficult to make any specific statements other than emphasize the importance of minimizing L_s.

ii. Amplifier Design Considerations: The major objective of this section is to review device technology. A brief but incomplete discussion on amplifier design is presented to indicate some important considerations. A major requirement for power amplifier design is to have a knowledge of the large-signal characteristics of the transistor [110]. Such characteristics can be obtained, for example, by using a load-pull technique preferably under computer control [111]. For satellite communications applications that require linear amplification with high efficiency, it is desirable to characterize the device by generating load impedance contours for constant output power, constant intermodulation distortion, and constant efficiency [86]. This procedure, described by Sechi [112], allows making the appropriate tradeoffs for specific applications.

Table III-17 describes the performance of an 8.5-W GaAs FET amplifier developed for RCA SATCOM by Dornan et al. [86] to indicate the excellent performance of GaAs FET amplifiers for communications applications.

c. Reliability

There is currently an intensive effort in many laboratories to improve the reliability of GaAs low-noise and power FETs and elucidate the failure mechanisms. The reliability of GaAs FETs is a function of the specific device design and

Table III-17. SUMMARY OF rf REQUIREMENTS AND PERFORMANCE FOR TWO ENGINEERING MODELS OF GaAs FET SATCOM AMPLIFIER [86]

Performance Characteristic	Specified Values	Engineering Models Unit #1	Unit #2
1. Type I	3.7–3.9 GHz		Type I
Type II	3.85–4.05 GHz		
Type III	4.0–4.2 GHz	Type III	
2. High-Power Operating-Point Power Output Minimums (Watts)	8.5 (EOL)	8.50	8.71
3. Gain at High-Power Operating Point (dB)	50–53	52.5–53.0	55.5
4. Small-Signal Gain (dB)	Typically 3–4 dB greater	55.0–56.0	59.0
5. Small-Signal-Gain Flatness (Any Channel) (dB p-p)	0.25	0.80	0.50
6. Small-Signal-Gain Slope (Any Channel) (dB/MHz)	±0.010	±0.02	±0.02
7. Small-Signal Gain Stability (Over Full Temp. Range) (dB p-p)	1.3	1.50	0.55
8. Level of IM Output Relative to Carrier Input Levels (dB):			
−3	−13	−13	−13
−10	−24	−23	−24
−17	−32	−34	−36
9. Harmonics Power Relative to Fundamental (Under Full Power) (dB)			
2nd Harmonic	−12	~60	~60
3rd Harmonic	−21	~70	~70
10. Input VSWR	1.25	1.23	1.25
Output VSWR	1.15	1.09	1.08
11. Group Delay (Any Channel):			
Variations ns p-p	0.5	0.25	0.83
Slope $(\Delta\tau/\Delta f)\eta$s/10 MHz	0.1	0.04	0.23
12. Phase Shift in Degrees for Input Drive Level of Carrier Below Operating Point of			
0 dB	22°	5.2°	20.0°
−3 dB	17°	8.0°	14.0°
−6 dB	14°	7.0°	11.0°
−10 dB	7°	4.0°	4.5°
13. AM/PM Conversion (°/dB)	2.0	1.40	1.70
14. AM/PM Transfer (°/dB) With Level of Unmodulated Carrier Relative to Modulated Carrier of:			
−20 dB	3.0	2.65	—
0 dB	2.0	1.50	—
15. Max. Dissipation Over Range of No Drive to Operating-Point Drive (Watts)	27	23.36	25.30
16. Max. Weight (lbs)	1.0	1.0	0.81
17. Max. Size* (in^3)	20	15.68	15.68

* Excluding output isolator.

geometry; we are therefore considering only the general features of the subject in this report.

The reliability of low-noise GaAs FETs has been extensively investigated [113-119]. The failure mechanisms and reliability are relatively well understood and it is claimed that the essential problems are practically solved [120]. The NEC-244 (Nippon Electric) low-noise GaAs FET is being used in the ANIK-B satellite that is operational and will be employed in the beacon and receiver sections of the RCA advanced SATCOM to be launched in October 1982. The Hewlett-Packard HP-2201 low-noise GaAs FET is being employed in the low-level stages of the solid-state power amplifier (SSPA) being developed by RCA for the advanced SATCOM.

The reliability of power transistors, however, is less well understood. A recent publication by Fukui et al. [120] describes the reliability of and failure mechanisms in GaAs power FETs developed by Bell Laboratories. The reliability aspects of power FETs are more complex than those of low-noise FETs. This is principally due to the fact that power devices operate at higher voltage (dc and rf) and current (dc and rf), and hence dissipate more power. The researchers at Bell Laboratories investigated GaAs power FETs with Al gates and silicon nitride passivation [120]. A total of 265 standard 6-mm-wide devices were aged under dc-bias conditions with and without rf drive at channel temperatures of 250, 210, and 175°C. One-million device hours were accumulated with no catastrophic failure. A very conservative estimate predicts that the failure rate for burnout at a maximum channel temperature in normal operation of 110°C would be below 100 FITs.* Degradation in the electrical parameters has been very slow even at 250°C channel temperature. It was estimated that the failure rate for gradual degradation at 110°C would be well below 100 FITs and most likely lower than 10 FITs. No deterioration in the properties of gates and ohmic contacts was observed. Diagnostic characterization has revealed that gradual degradation in the sample devices is caused by deterioration in the channel material. There was no noticeable difference in gradual degradation between devices aged with and without rf drive at the same channel temperature for more than 3000 h.

RCA has carried out a study of GaAs power FETs for communications satellite applications [86,121]. The design operating life of the satellite is 10

*One (1) FIT is one failure in test in 10^9 device-hours.

years and thus requires highly reliable devices. The influence of parameter screening and device operating conditions on the lifetime of power GaAs FETs has been investigated. Under rf operating conditions, power FETs are subject to wide voltage and current excursions, which stress the devices near to maximum values. To ensure reliable, long-life operation, effective device screens and operating design guidelines were established.

RCA has conducted rf accelerated temperature testing of both commercial and screened GaAs FETs. The commercial devices were of two types: Fujitsu FLC-30 MM and FLS-50 MM. Both FETs are of similar construction with the FLC-50 having approximately 40% more active area. Eight devices of each type were tested at a channel temperature of 200°C with an rf input drive of 30.0 dBm at 4.0 GHz.

The commercial FLC-30 devices exhibited an S-shaped failure distribution with excessive gate current under rf drive as a precursor to early failure. Drain-source electromigration with Au-Ga hillock growth was the predominant failure mechanism. The FLS-50 devices, effectively being driven at a 30% lower rf power density, showed less severe electromigration at equivalent test times.

As a result of the testing of these commercial FETs, a screening and qualification procedure for "flight-quality" FETs was devised. Nineteen FLC-30 MA FETs that comply with this qualification procedure were then tested at a channel temperature of 190°C. An additional 10 qualified devices were tested at a channel temperature of 215°C. These FLC-30 MA devices were tested at a drain-source voltage of 9.0 V and an input drive of 27.5 dBm at 4.0 GHz.

These 29 devices had no infant failures. The predominant failure mode was graceful in nature; the predominant mechanism was drain-source migration with evidence of gate voiding in some devices. A catastrophic failure mechanism associated with hillock growth was identified. This effect occurs well after the useful life of the device, which is defined by a 0.3-dB decrease in the output power of the device.

Device screening, with a specification that includes a high-temperature rf burn-in, has been shown to eliminate early failures in GaAs power FETs. Flight-quality devices screened to this procedure are estimated to have an average failure rate, over a 10-year mission, of 2.0 FITs. This estimate is for an operating channel temperature of 85°C.

It is clear that while more research on GaAs FET reliability must be carried out, the preliminary results are very encouraging. RCA expects to launch a SATCOM satellite with 8.5-W GaAs FET power amplifiers in October 1982. Table III-18 lists some examples of space-qualified GaAs FETs.

Table III-18. SPACE-QUALIFIED GaAs FETs*

Device Type	Manufacturer	Part No.	Comments
Low-Noise	Nippon-Electric, Japan	NEC-244	1. Being used in ANIK-B, operational
			2. Will be used in RCA Advanced SATCOM, October 1982
			3. Al-Gate Device
Low-Noise	Hewlett-Packard Palo Alto, Calif.	HP-2201	1. Will be used in driver section of SSPA for RCA Advanced SATCOM
	Fujitsu, Japan	FLC-02	2. Al-Gate Device
Power-FET	Fujitsu, Japan	FLC-08	1. Used in SSPA, RCA Advanced SATCOM
		FLC-15	
		FLC-30	2. Al-Gate Device

*Illustrative examples.

C. MULTIGIGABIT-RATE GaAs DIGITAL INTEGRATED CIRCUITS

The state of the art of GaAs MESFET-based digital ICs has come a long way since the publication of Van Tuyl and Liechti's pioneering paper in 1974 [122]. The circuit configuration they described, since termed buffered FET logic (BFL), is still one of the most popular among laboratories all over the world. Before discussing the features of GaAs digital ICs and their relative merits compared with Si technology, we present the following background information:

(a) Liechti et al. of Hewlett-Packard have developed a GaAs BFL three-chip set consisting of an 8:1 parallel-to-serial converter (programmable word generator) that generates 8-bit words at 5 Gb/s; a pseudorandom bit-sequence generator (PRBS) that produces a maximum-length sequence of 1023 bits at 2.5 Gb/s; and a recurrence-relationship test chip that receives and checks the PRBS for errors at 2 Gb/s [123]. They report a 30% functional yield for these MSI circuits consisting of about 500 active devices per chip [123]. The FETs used have 1 μm-gate lengths.

(b) Eden et al. of Rockwell International have reported on an 8x8 multiplier using the Schottky-diode FET logic (SDFL) approach that they pioneered [124]. The multiplier comprises 1008 SDFL gates with 6048 devices. This chip represents the highest level of integration reported to date for GaAs ICs. The multiplier has a multiply-time of 6 ns and dissipates 0.8 W. This performance compares favorably with a state-of-the-art Si ECL multiplier (TRW) that has a multiply time of 70 ns and a power dissipation of 1.2 W.

(c) Researchers at Thomson-CSF have described a quasi-normally-off GaAs FET logic circuit called the low-pinchoff voltage logic (LPFL). An 8-bit static random-access memory with 6000-μm^2 cells and devices with 0.8-μm features with a decode time of 0.6 ns has been reported with LPFL [125].

The dramatic development of GaAs digital ICs in less than a decade clearly attests to their potential. The viability of GaAs-FET-based logic, the integration levels that can be achieved, and the potential advantages compared to Si logic technology are still being debated. A dispassionate comparison of the potential of GaAs and Si technology is not available in the literature. Such a comparison is difficult because of the many variables involved. Since the major demand for digital ICs is at lower speeds, Si technology has really not been pushed to its ultimate limits insofar as speed is concerned. The current effort on the development of Si-based very high speed integrated-circuit (VHSIC) technology will go a long way toward answering many of the questions raised above. The situation is further complicated by the fact that functional speed in logic subsystems can be increased by parallel processing - the advanced state of the art of Si technology (e.g., VLSI capability) is particularly suited for the parallel processing approach.

In spite of this uncertainty, it is clear that GaAs digital technology will have an impact on some important specialized applications, in particular, ultrahigh-speed real-time signal processing. We can make the following general statements:

(1) Computer simulation of GaAs and Si MESFETs with comparable geometries (1- and 0.5-μm gate lengths) shows that the GaAs MESFET can switch twice as much current as its Si counterpart in the same time [126]. This fact is a consequence of the higher low-field mobility of GaAs. Since the gate-source capacitance for GaAs and Si MESFETs is comparable, the GaAs switch is about twice as fast at the Si device. This is independent of fan-in and fan-out.

(These simulations do not include nonequilibrium effects such as velocity overshoot. Velocity overshoot will tend to favor the GaAs gate [92]).

(2) Measured switching data of Greiling et al. [127] show that GaAs gates are two-to-six times faster than comparable Si gates depending upon bias conditions. At low-bias conditions, the higher peak velocity in GaAs results in the six-times advantage and, at high-bias voltage, comparable values of saturated velocity reduce the advantage to a factor of 2 [126].

(3) Turn-on and turn-off time are two-to-three times lower (faster) for GaAs MESFETs than for comparable Si MESFETs [126]. Table III-19, taken from reference 126, shows a comparison between 1- and 0.5-μm GaAs and Si MESFETs.

Table III-19. TWO-INPUT MESFET NOR GATE

(Turn-on time (τ_{ON}), turn-off time (τ_{OFF}), average switching delay (τ_d), average power (P_D), and power delay product ($P_D\tau_d$) for gates using 1- and 0.5-μm Si and GaAs MESFETs. F denotes fan-in, and E fan-out) [126].

Gate	F/E	τ_{ON} (ps)	τ_{OFF} (ps)	τ_d (ps)	P_D (mW)	$P_D\tau_d$ (pJ)
1 μm Si	1/1	131	315	223	5.7	1.3
1 μm GaAs	1/1	55	101	78	12.3	0.96
1 μm Si	2/2	186	449	318	5.7	1.8
1 μm GaAs	2/2	84	152	118	12.3	1.5
0.5 μm Si	1/1	92	283	188	8.9	1.7
0.5 μm GaAs	1/1	47	110	79	14.7	1.2

(4) Barna has computed the propagation delay as a function of interconnection capacitance in two-input NOR gates with a fan-out of 2 for Si NMOS, CMOS, ECL, and GaAs depletion and enhancement MESFET technologies [128]. Figure III-28 shows these computations for 1-μm minimum geometries at gate power dissipation levels of 1 mW (typical LSI dissipation - 2000 gates/chip) and 0.2 mW (typical VLSI dissipation - 10,000 gates/chip). Clearly, under LSI conditions, GaAs technologies provide a performance superior to that of Si for custom logic. Under VLSI conditions, interconnection delay becomes more critical, and the advantage of GaAs technology is less clear. Eden has pointed

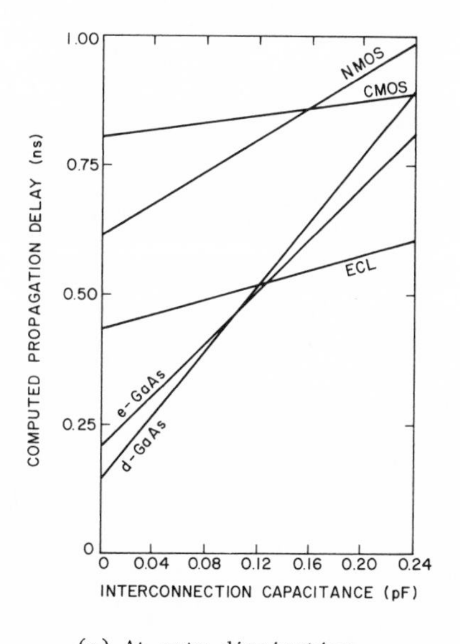

(a) At gate dissipation, P_D = 1 mW (typical LSI).

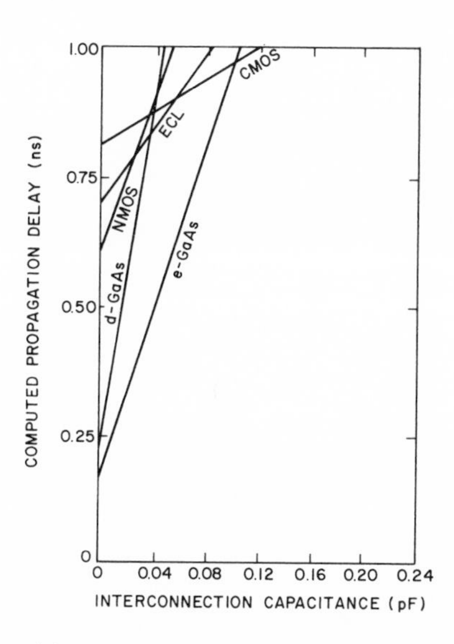

(b) At P_D = 0.2 mW (typical VLSI).

Figure III-28. Propagation delay vs interconnection capacitance for various technologies; 2-input NOR gate, fan-out = 2, gate length = 1 μm [128]. Redrawn with permission of VLSI DESIGN Magazine.

out, however, that a VLSI chip is designed so that the interconnection lengths in critical areas are small; long interconnections are used in less critical regions where slower signal propagation can be tolerated [129]. This consideration must also be factored in.

It is reasonable to conclude at this stage of development that while GaAs is indeed suitable for VLSI with 10,000 gates/chip, it is likely to have its greatest impact on specialized custom logic. The impact on VLSI circuits built with library cells and automatic placement routines will probably be limited [128].

1. GaAs MESFET-Based Logic Technologies

We will review the major GaAs MESFET-based logic technologies under development [130]. The superior electron dynamics of GaAs, along with the availability of semi-insulating (10^7-10^8 Ω·cm) GaAs substrates, make it very attractive for multigigabit-rate digital integrated circuits. The electron mobility in GaAs is about six times that of similarly doped bulk Si. Furthermore, the electron mobility in typical GaAs MESFET channels is 10 times that of Si NMOS FETs under strong inversion. Electron drift velocities in typical 1-μm gate GaAs MESFET channels (~1.5-2x10^7 cm s^{-1}) are several times those in Si NMOS devices (~0.65x10^7 cm s^{-1}) and occur at lower voltages [130]. Eden et al. have compared "K" values defined by the equation $I_{DS} = K(V_{gs} - V_p)^2$ for GaAs and Si devices (fig. III-29) [130]. The K values for GaAs MESFETs for gate voltage (V_{gs}) close to threshold (V_p) are higher than those for Si FETs. This implies that for identical loading and interconnect capacitances, GaAs MESFETs can achieve higher switching speeds at lower voltage swings. (This is identica to our previous statement that GaAs MESFETs are better current switches.) The semi-insulating GaAs substrate also results in much lower interconnect and parasitic capacitances than does bulk Si ICs (comparable to silicon on sapphire

For a viable digital IC technology, the following requirements must be met simultaneously:

i. Ultrahigh switching speed. This implies low propagation delay, τ_d, and fast rise and fall times.
ii. Low power dissipation per gate, P_D.
iii. Low dynamic switching energy; i.e., low power-delay product, $P_D\tau_d$.
iv. High process yield.
v. High gate density.

Figures are not given, since these requirements are interrelated. Delay, power dissipation, and gate densities must be traded off, depending on specific requirements.

a. Buffered FET Logic (BFL)

The first GaAs MESFET logic IC reported used the buffered FET configuration [122]. The BFL configuration is perhaps the most popular configuration and is being pursued by HP, RCA, Hughes Research Laboratories, Texas Instruments, Thomson CSF, Plessey, Philips LPE, Fujitsu, NEC, and others [125]. BFL circuits represent the fastest GaAs ICs reported to date (e.g., 5 GHz 8:1

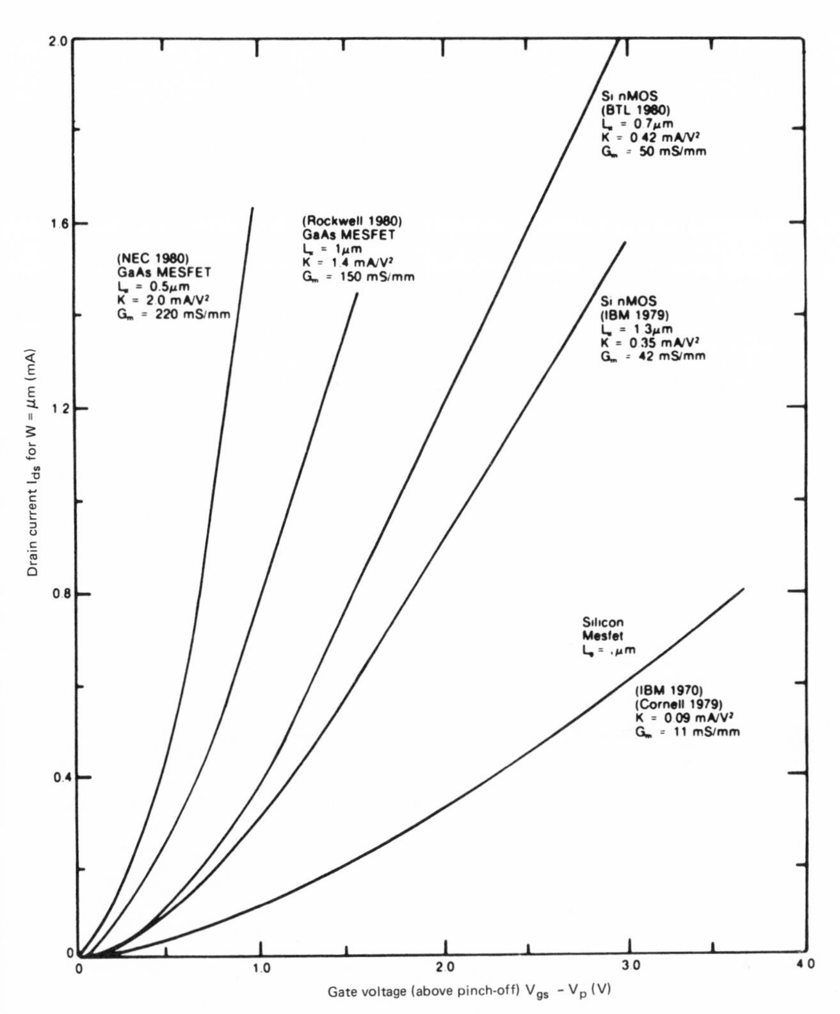

Figure III-29. Comparison of characteristics of GaAs silicon and short-channel MESFETs, and silicon NMOS drain current vs gate voltage (above threshold). In the MESFETs, the effective source-drain channel length equals the gate length, whereas in the NMOS devices it is shorter (as low as 0.3 μm for the L_g = 0.7 μm NMOS devices). GaAs HEMT devices have even sharper turn-on characteristics, and achieve G_m = 409 mS/mm transconductances with L_g = 1.7 μm (at 77 K)[128]. Reprinted with permission of VLSI DESIGN Magazine.

multiplexer [123]). This configuration is on the high-speed side of the speed-power spectrum and is most suited for MSI and perhaps LSI levels.

Figure III-30 shows the basic inverter and NAND/NOR BFL circuit configuration. The characteristic features of the BFL configuration are as follows:

(1) Circuits typically employ depletion-mode GaAs MESFETs with $-3\ V < V_p < -1\ V$. The gate requires a negative voltage to turn off the n-channel, and a positive drain voltage must be employed; level shifting must therefore be provided to make the input and output compatible.

(2) In BFL, negative-voltage swing operation is achieved by level-shiftin the positive drain voltage. This is done by using several forward-biased diodes in the source follower/load driver circuit. The drawback of this configuration is that a significant fraction of power is dissipated in the level-shift circuits.

(3) The nonlinearity of the MESFET is used to implement the logic function. Higher-level logic, i.e., multiple NAND/NOR functions are possible by series/parallel combinations of switch transistors. Note that multiple input NAND/NOR gates can use one load driver and level-shift chain. The load driver (or buffer) transistor can be eliminated in noncritical paths, resulting in reduced power dissipation with a corresponding increase in delay [130].

(4) A positive ($+V_{DD}$) and negative ($-V_{SS}$) power supply are required.

The attractiveness of the BFL approach lies in these features:

(1) The BFL approach can be implemented using a single epitaxial or ion-implanted ($\sim 10^{17}\ cm^{-3}$, 0.15-0.2 μm) layer by a process similar to that of a discrete microwave MESFET. Device isolation can be achieved by mesa isolation, by proton or B implant, or by single selective implant.

(2) Multiple implantation, i.e., n^+ layers for contacts, can be added to improve performance.

(3) The higher speed of this configuration makes possible MSI circuits (such as multiplexers/demultiplexers, PRN generators, and programmable dividers) that can find immediate application. These circuits can be used for "front end" signal processing to extend the frequency range of instrumentation, for increasing data bus rate by multiplexing/demultiplexing, in frequency synthesizer applications, etc.

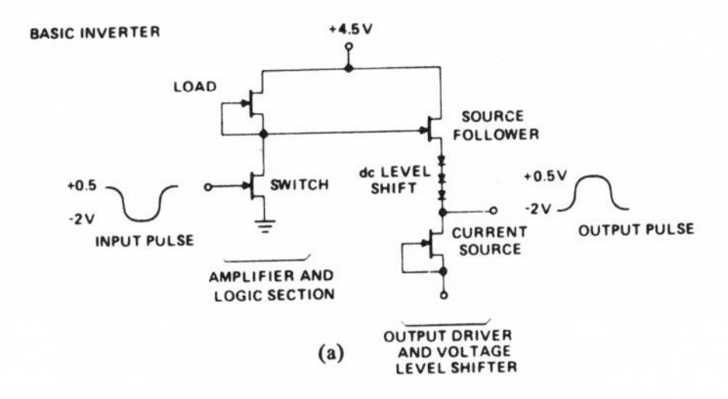

(a) Basic inverter circuit.

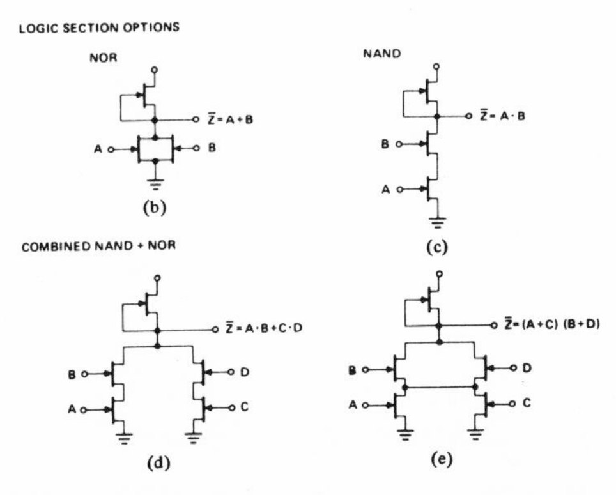

(b)-(e) Options for the input section for NOR, NAND, and combined NAND-NOR functions.

Figure III-30. Basic circuit configurations for buffered FET logic [76]. © 1982 IEEE.

b. Schottky-Diode FET Logic (SDFL)

The SDFL is a variant of the BFL approach pioneered by investigators at Rockwell International [130]. In SDFL, the diodes do the logical positive ORing at the input, and depletion mode MESFETs provide inversion and gain.

Figure III-31 shows some typical SDFL gate configurations. The SDFL configuration has the following characteristics:

(1) It uses clusters of small (typically 1 μm x 2 μm) Schottky diodes to perform logic OR function. These are further processed by MESFETs for series NANDing, drain dotting, etc.

(2) High fan-in can be achieved at the first (diode) level. At the second (MESFET) level, fan-in is restricted to 2 or 3 in a manner similar to BFL.

(3) Moving the diodes from a high-current to a low-current path results in substantial saving in power dissipation.

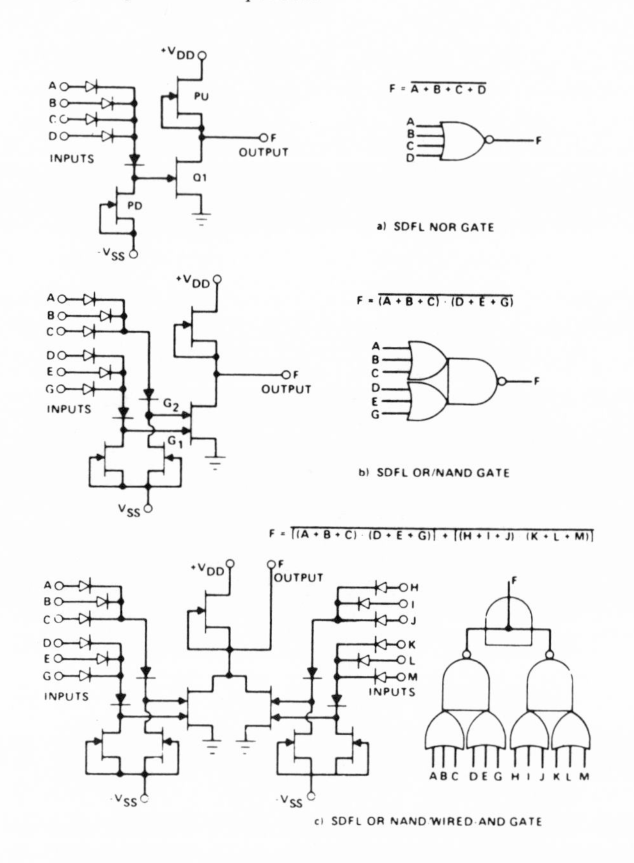

Figure III-31. Comparison of 1-, 2- and 3-level SDFL gate configurations. All FETs are depletion-mode, typically $-1.5 < V_D < -0.5$ V; unshaded diodes are very small high-speed-switching Schottky diodes; shaded diodes are larger-area, higher-capacitance voltage-shifting diodes [76]. © 1982 IEEE.

(4) The use of small 1-μm x 2-μm diodes instead of MESFETs results in a saving in circuit area, leading to higher packing density. Use of two-terminal diodes instead of the three-terminal MESFETs reduces the number of vias (or overcrossings).

(5) As for the BFL configuration, a positive (V_{DD}) and negative ($-V_{SS}$) power supply are required.

As indicated earlier, the SDFL approach yields the highest complexity yet achieved with GaAs ICs [130]. This is an 8x8 multiplier with 1008 SDFL gates comprising 6048 active devices [130]. The SDFL configuration may be more suited to LSI and VLSI levels of integration than the BFL architecture. It has been stated that BFL circuits are at least 20% faster under similar conditions of power consumption and loading [130]. While this statement appears reasonable, it must not be taken as definitive.

The advantages of SDFL accrue at the expense of increased fabrication complexity. The increased fabrication complexity, however, results in a planar "silicon"-like process. The key features are (a) that multiple localized ion implantation (selective implantation) is used to define the MESFET channel, diode active layer, and MESFET and diode ohmic contact regions; these implantations are carried out through dielectric (Si_3N_4) or photoresist masks [130]; (b) that the switching diodes are very small, usually 1 μm x 2 μm in size. This has necessitated improvement in lift-off technology such as the development of dielectric assisted lift-off, or multilevel lift-off processes [130]. The SDFL gate schematic cross section is shown in figure III-32. Fabrication processes used are reviewed in many papers [130,131] and will not be discussed.

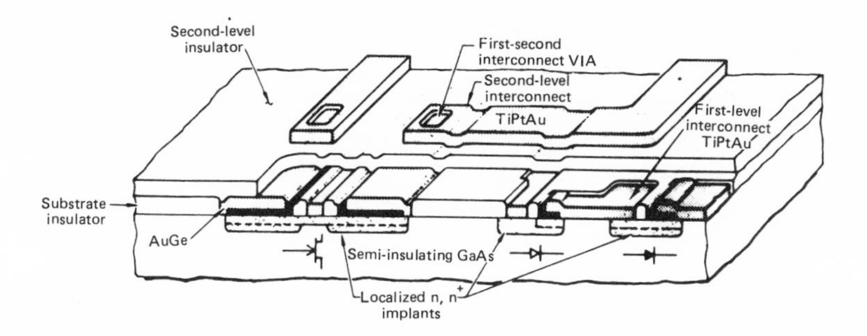

Figure III-32. Cutaway view of a planar GaAs IC showing a dual-gate FET, a diode, and interconnects [76]. © 1982 IEEE.

c. Enhancement-Mode FET Logic

Enhancement-mode MESFETs (e-MESFETs) offer the following advantages:

(1) The devices are normally-off, and gates can be coupled directly without any level shifting, leading to simpler circuits.

(2) Circuits require only one power supply in contrast to the two required for d-MESFETs; this, again, leads to lower circuit complexity.

(3) The power consumption in e-MESFET logic circuits is lower than for d-MESFETs; i.e., the former operate in the low-power region of the power-speed spectrum.

These advantages must be traded off against the following:

(a) The permissible voltage swing is low since Schottky-barrier gates on GaAs cannot be biased above about 0.7 V without drawing significant current. This in itself is not a significant disadvantage since a 0.5-V internal logic swing is desirable in ultralow power ICs. This implies, however, that an extremely tight control must be maintained on the doping and thickness of the relatively thin (~0.1 μm) active layer to keep it depleted under zero applied gate bias and still obtain good transconductance in the ON state. For reasonable noise margins and dynamic performance, the MESFET pinchoff voltages must have standard deviations better than about 25 mV - a very stringent technological constraint requiring excellent control over SI GaAs substrate quality, ion implantation, and annealing.

(b) The situation can be alleviated somewhat by using a p^+n junction gate (e-JFET) [132] since GaAs p^+n junctions have built-in voltages of about 1-1.2 V. The p^+n junction gate can be achieved by an added ion implantation step [132]. This technology is still under development.

(c) The use of a heterojunction gate (e.g., p-type $Ga_{0.5}Al_{0.5}As$) can increase the barrier potential to 1.4 V [133]. This requires heteroepitaxy and more complex fabrication.

(d) The e-FET logic circuits are more sensitive to fan-in (FI) and fan-out (FO), and a significant increase in propagation delay must be expected for reasonably complex logic circuits (i.e., complex compared to simple ring oscillators).

In 1977, Fujitsu reported propagation delays of 280 ps with 265 μW/gate on ring oscillators with 1.2-μm e-MESFETs [134]. Thomson CSF obtained 650 ps with 20 μW/gate [135], and NTT and Hughes have reported 77 ps with 0.98 mW/gate [136], and 72 ps with 0.89 mW/gate [137], respectively.

Figure III-33 a shows the Directly Coupled FET Logic (DCFL) gate. The normally-off FET starts conducting for positive gate voltages. Logic "0" corresponds to near 0 V, and logic "1" is a positive voltage that is capable of turning on the following normally-off FET. The value of logic 1 is limited by the onset of gate forward conduction. Figure III-33(b) shows a normally-off FET with a depletion FET load. This well-known configuration will sharpen the transfer characteristic and improve the speed and power/delay product by about a factor of 2. The depletion-mode active load will, however, require a different carrier profile. This implies the use of a technology such as localized multiple implantation.

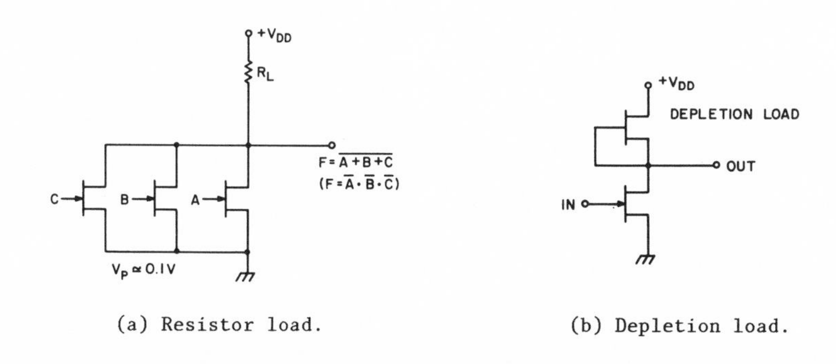

(a) Resistor load. (b) Depletion load.

Figure III-33. Enhancement-mode MESFET or JFET-DCFL circuit.

From a static point of view, the fan-out capability of the directly coupled FET logic is excellent, since it is determined by the very low gate-leakage currents. However, from a dynamic point of view, the switching speeds are reduced by the gate capacitance loadings by a factor of approximately 1/N where N is the number of loading gates, as in silicon MOS. In general, the current through the resistor, R_L, or active load is kept fairly low in DCFL so as to reduce static power and improve noise margin by reducing the "on"-voltage drop of the FET (output "low" voltage). Consequently, the output rise time under heavy fan-out loading conditions is poor.

Implementation of a MOSFET or MISFET technology in GaAs would eliminate the logic swing limitation completely, but attaining such devices has proven difficult. Some simple ring oscillators have been fabricated with a directly coupled FET logic implemented with buried-channel GaAs MOSFETs with resistor loads. However, stable oxides have not yet been achieved in such circuits, and gate thresholds may shift according to the prior input-signal history. This instability does not prevent the demonstration of ring oscillators and other simple circuits in which the input wave form has a precisely symmetrical (50% duty cycle) nature, but severely degrades performance of general digital circuits. There is also a fundamental limitation that has prevented progress in GaAs MOS/MIS technology. This limitation is that the Fermi level in GaAs appears to be pinned at near midgap, minimizing the variation in surface potential that can be achieved by an insulating gate [138]. Thus inversion mode, normally-off GaAs MOS/MIS devices have not been fabricated. We will discuss InP and $Ga_{0.47}In_{0.50}As$ MISFETs in Section III.E (Exploratory Development).

d. Quasi-Normally-Off or Low-Pinchoff FET Logic (LPFL)

The concept of quasi-normally-off MESFET logic described by researchers from Thomson CSF [139] derives from practical considerations. The objective is to achieve single-power supply operation while relaxing the stringent restrictions on pinchoff voltage uniformity required by e-MESFET logic circuits.

Figure III-34 shows the circuit diagram of an LPFL circuit [139]. The switch transistor T_1 is not quite normally-off, as shown in the inset. The analysis presented in reference 139 shows that a variation in V_p of T_1 that can be tolerated is twice that for e-MESFET circuits. Furthermore, it shows that higher logic swings (of about 0.8 V compared to about 0.6 V) can be achieved than for DCFL. The advantages claimed for LPFL are:

(1) High potential fabrication yield due to a relatively large tolerance to material and/or fabrication nonuniformities.

(2) Single power supply (∿3 V) operation leading to packing densities of about 400 gates/mm^2.

(3) Attractive speed-power tradeoff with propagation delays in the 100- to 200-ps range and a power-delay product of 50-200 fJ.

2. Performance

We will summarize the speed-power performance of the various GaAs MESFET logic technologies. The performance data presented are under laboratory conditions and use fabrication technologies ranging from single active layer to multiply-implanted wafers, from conventional contact photolithography to

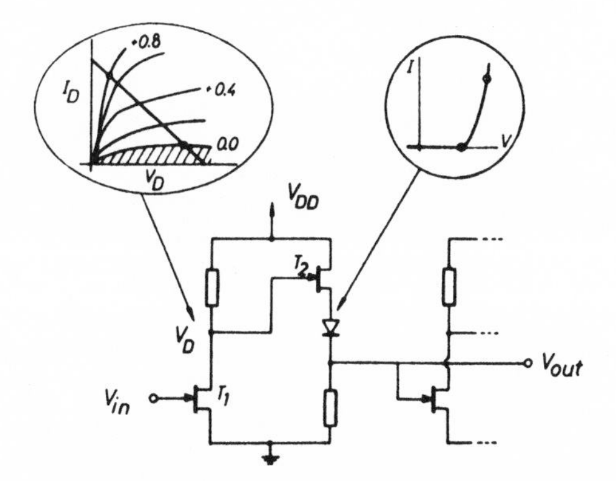

Figure III-34. Quasi-normally-off (LPEL) inverter [139].
© 1980 IEEE.

direct-write-e-beam lithography. The goal is to provide a broad perspective, and better data may be found in current literature.

Figure III-35 shows a gate propagation delay vs power dissipation chart of various GaAs and Si logic technologies [140]. This chart is based on ring-oscillator data. While a ring oscillator is a very convenient circuit for evaluating a logic technology, the data should be interpreted with caution. Ring oscillator data may be optimistic because of unrealistically low loading capacitances or operation in quasi-Class A conditions - i.e., at voltages below that required to provide sufficient noise margin in practical digital circuits. Another example of misleading ring oscillator data is that excellent results were obtained on ROs fabricated with GaAs MOSFETs [141]. While ROs with symmetrical waveform operation (50% duty) can be realized, the drift in oxide-gate characteristics does not allow this device to operate in digital combinatorial circuits. Figure III-35 also includes data on high-speed Si NMOS devices. Table III-20 lists the best (Feb. 1982) data published in the literature.

The Si NMOS data listed in table III-20 are impressive (e.g., a 65-ps propagation delay for a 0.25-μm gate-length device). This small gate length, however, requires very complex technology such as x-ray and/or e-beam lithography.

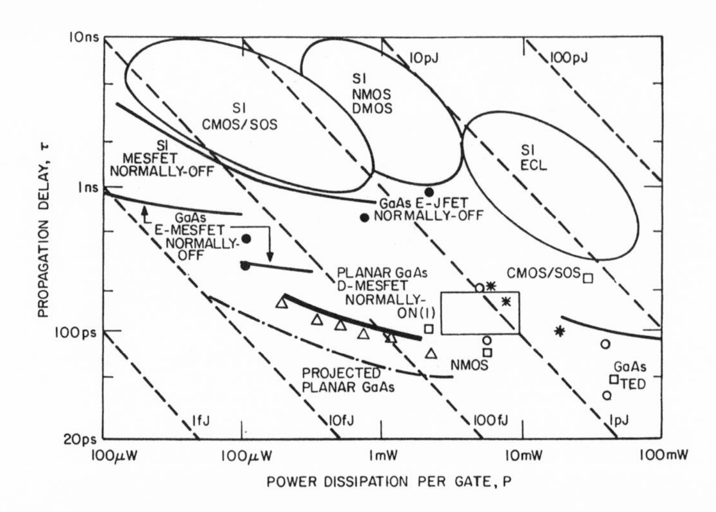

Figure III-35. Propagation delay-dissipation chart for modern-logic technologies (~1981).

For comparable geometries, the GaAs devices have superior performance. The excellent performance of 0.25 μm NMOS Si devices illustrates why it is very difficult to make predictions about the commercial viability of GaAs ICs for digital logic. Si ICs are leveraged strongly by the investment in the Si technology base and the vast market for low-speed ICs. It can be argued, however, that advances in submicrometer lithography can also be applied to GaA technology and that GaAs devices will always be faster than their counterparts

Table III-20 also shows data for a heterojunction FET (HEMT) at low temperature (77 K). Note that logic speeds approach those for Josephson junction (JJ) devices, but operate at 77 K. Refrigeration required is thus less complex than that required for JJ devices. HEMTs are discussed in greate detail in Section III.E.

Table III-21 shows some published data on flip-flops that have been demon strated with various GaAs IC approaches. Note that the maximum toggle frequen of the flip-flop depends on the specific circuit architecture chosen. The equ valent propagation delay computed from flip-flop performance is also shown.

TABLE III-20. RING OSCILLATORS

GaAs IC Configuration	Gate Length (μm)	Gate Width (μm)	Power Dissipation (mW)	Propagation Delay (ps)	Speed-Power Produce (pJ)	Fan-In	Fan-Out	Comments
1. GaAs BFL [123]	1	?	10	50	0.5	1	1	
2. GaAs BFL [112]	0.5	10	5.6	83	0.46	1	1	
3. GaAs SDFL [124]	1.0	5	0.17	156	0.03	2	2	
			0.62	87	0.05	2	1	
4. GaAs DCFL [136]	0.6	20	1.9	30	0.06	1	1	
(e-MESFETs)	0.6	20	0.05	78	0.004	1	1	
5. GaAs LPFL [139] (Quasi-Normally-Off)	1.0	35	1.2	100	0.12	1	1	
6. GaAs/GaAlAs HEMT [143]	1.7	?		17		1	1	77 K operation
7. Si NMOS [143]								
a.	1.0	?	0.65	230	0.15	1	1	Note effect of increasing fan-in and fan-out
b.	1.0	?	0.2	1000	0.2	3	3	
8. Si NMOS [143]	0.25	?	1.7	30	0.05	1	1	

TABLE III-21. GaAs FET FREQUENCY DIVIDERS

GaAs IC Technology	Circuit Approach	Theoretical Toggle Frequency	Measured Toggle Frequency (GHz)	Equivalent Delay (ps)	Power Dissipation/ Gate (mW)	$P_D \tau_D$ (pJ)
1. GaAs BFL 1 μm [123]	NAND/NOR ÷ 2, complementary clock	1/2 τ_D	6.5	78	10	0.78
2. GaAs BFL 0.7 μm [130]	D-flip-flop ÷ 2	1/5 τ_D	3	67	40	2.7
3. GaAs SDFL 1 μm [130]	D-flip-flop ÷ 2	1/5 τ_D	1.9	105	2.5	0.26
4. GaAs DCFL 0.6 μm [136]	D-flip-flop ÷ 8	1/4 τ_D	3.8	66	1.2	0.08

These propagation delays are in reasonable agreement with those computed from RO data.

3. Projections

The summary of our discussion (below), and an attempt to project developments in the state of the art, are confined to conventional GaAs digital IC technology. It is clear that GaAs offers major performance advantages over Si for multigigabit-rate digital ICs. The high mobility and electron velocity in short-channel (≤1 μm) GaAs MESFETs result in much higher transconductance (or drain currents) than may be obtained for Si n-channel FETs at equivalent gate biases. This improved performance, along with the availability of the SI GaAs substrate, makes possible higher switching speeds and lower power-delay products than for Si devices. Experimental data on simple ring oscillators and more complex sequential logic circuits confirm this contention. Circuits with LSI complexity have been demonstrated with GaAs in a laboratory environment. Functional yields as high as 30% at 500 gate complexity have been reported.

Until recently, the technology of GaAs was fairly primitive, especially when compared to Si technology. This situation is changing rapidly with the development of multiple localized implantation technology, various modern epitaxial techniques such as MBE and MOCVD, and the increased availability of SI GaAs substrates suitable for direct implantation. The development of multiple localized implantation, in particular, has had a dramatic impact on the realization of planar "Si-like" ICs of LSI complexity in GaAs. Figure III-36 illustrates the evolution of GaAs IC technology [76].

We can envision the following scenario:

(1) GaAs MSI circuits fabricated by use of conventional contact and/or projection optical lithography with 1-μm features will become available for front-end high-speed digital systems. This category includes programmable dividers (prescalers), multiplexer-demultiplexers, PRN generators, etc. The three-chip set described by Hewlett-Packard falls in this area and will extend the range of instrumentation of these devices. The use of multiplexers/ demultiplexers in conjunction with high-speed Si logic circuits can lead to data busses operating at multigigabit rates.

(2) GaAs e-MESFET-based memory circuits using 1-μm feature size are under development in Japan. These will be suitable for high-speed cache memory applications.

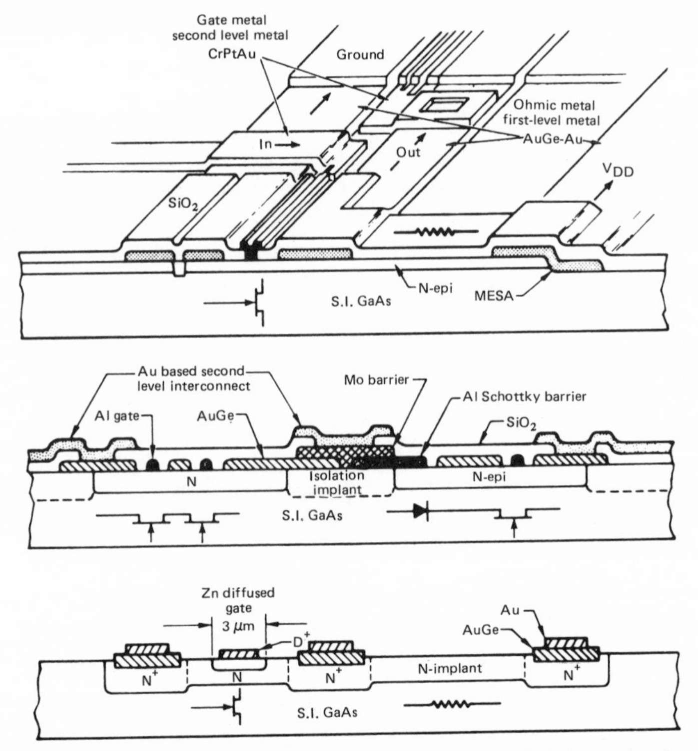

Figure III-36. Schematics showing the evolution of GaAs technology from the simple mesa-epitaxial D-MESFET to the present sophisticated planar multiply-implanted GaAs IC [76]. © 1982 IEEE.

(3) Speed-power optimized GaAs LSI circuits like the 8x8 multiplexer developed by Rockwell International will extend the range of fast-Fourier transformer (FET) processors and digital filters (formation of sum of products).

(4) The improvement in modern lithography (x ray, direct e-beam) that is being leveraged by Si technology will be applied to GaAs ICs. GaAs ICs with submicrometer gates with propagation delays in the 10- to 20-ps range will be developed by the mid-to-late 1980s.

The major applications foreseen for GaAs digital ICs are in front-end critical applications, real-time signal processing for military systems, and special high-speed circuits for space communications.

D. MICROWAVE INTEGRATED CIRCUITS

This section briefly discusses monolithic (MMIC) and hybrid microwave integrated circuits (HMIC) using GaAs. MMIC design considerations are reviewed by Pucel [144], and Belohoubek has reviewed trends in both MMICs and HMICs for EW applications [145]. It is emphasized at the very outset that we view these two approaches as complementary technologies.

The term monolithic implies that all active and passive elements including interconnections are formed into the bulk or defined on the surface of an insulating (or semi-insulating) substrate. The potential advantages of the MMIC approach are:

(1) Low cost: This advantage can be realized if a large number of identical circuits is required. The potential low cost is due to the use of batch fabrication processes (as in the case of digital circuits) and the minimization of wire bonds within the chip. Wire bonding is a labor-intensive operation, and the wire-bond inductance can vary from circuit to circuit, leading to unacceptable variations in performance.

(2) Improved performance: The elimination of wire bonds, and embedding all active and passive components within a printed circuit format, minimize parasitic elements (reactances) that limit performance. This is particularly true in broadband applications and in applications (e.g., at 20 GHz and higher) where one cannot tolerate any loss in device intrinsic performance.

(3) Improved reproducibility: This is a well-known consequence of the use of lithography to define devices and circuits. Material uniformity and other technological constraints must, however, be borne in mind.

(4) Small size and weight.

(5) Circuit design flexibility.

These potential advantages have to be balanced against the following disadvantages or problems:

(1) Low active-device/passive-component area ratio: In general in MMICs, only a small fraction of the chip is occupied by active devices. This is particularly true at the lower (10 GHz) microwave frequencies. In addition to poor usage of wafer real estate, the high processing cost and lower yield of

active device fabrication apply to the larger area occupied by passive circuitry [144]. This disadvantage is absent in the hybrid approach, in which the active- and passive-component technologies are separate and independent of each other.

(2) Circuit readjustment (tweaking) and trouble-shooting difficulty: The small chip size and complete on-chip interconnection make circuit fine-tuning and trouble-shooting difficult. This is an example of a feature that can be either an advantage or disadvantage depending on the circumstances. If the design is good and yields are high, this feature results in a cost-effective circuit. The development of CAD techniques will allow the design and extensive simulation of the circuit prior to fabrication. Parameter sensitivity analyses must be carried out to determine an optimal tradeoff between performance, sensitivity to material parameter variations, and sensitivity to fabrication tolerances. The development of microwave probes that allow on-wafer component testing will aid in trouble-shooting [146].

(3) Crosstalk: Unwanted rf coupling between adjacent lines on a circuit can occur if they are in close proximity [144]. This crosstalk can cause positive feedback in high-gain chains and lead to oscillations. A line spacing (S) greater than three times the substrate thickness (H) is recommended as a rule of thumb [144]. This factor increases chip area by decreasing packing density.

(4) Medium power circuits: MMICs are more suited to low-power level rather than medium-power (>1 W) applications. The thermal and electrical design considerations are, in general, not compatible. To efficiently remove heat by proper heat-sinking, it is necessary to use relatively thin (<100 μm) substrates to decrease thermal resistance. Use of thin substrates increases circuit losses and degrades circuit Q and performance. Good heat-sinking also requires that the bottom of the substrate be metallized and bonded to a metal with good thermal conductivity. This metal is generally at ground potential and introduces parasitic capacitances to ground. The presence of metal (ground plane) also influences the design of circuit elements such as inductors. One design strategy is to thin the substrate only under the heat-generating active device and filling that with metal (via-hole heat-sinking) [144]. While this approach may result in satisfactory performance, the differences in thermal expansion between the metal and substrate may lead to reliability problems.

In summary, the best candidates for monolithic applications are low-power circuits that have a very large volume potential and operate at frequencies of X-band or higher. Table III-22 summarizes the potential of MMICs. Note that in many critical applications, monolithic functional blocks used in hybrid circuits may result in improved performance. Examples are monolithic multi-stage amplifier "gain" modules, GaAs MESFETs with partial impedance matching, and mixer-LO blocks.

Table III-22. MONOLITHIC MICROWAVE INTEGRATED CIRCUITS (MMICs)

Potential Advantages

1. Low Cost: when a very large number of identical circuits is required
2. Improved Performance: elimination of wire-bond parasitics improves performance, particularly bandwidth
3. Small Size: particularly at high microwave frequencies (>10 GHz)
4. Improved Reproducibility and Reliability

Potential Disadvantages

1. Low Active-Device/Passive-Component Area Ratio
2. Circuit Adjustment and Trouble-Shooting Difficult
3. Crosstalk Problems Limit Packing Density
4. Problems with Integration of Medium-Power (>1 W) Devices

Potential Applications

1. Low-Power (<1 W), Very Large Volume Applications at X-Band and Higher Frequencies
   Driving Force - Potential Low Cost
2. Functional Blocks in Critical Applications
   Driving Force - Improved Performance

1. Hybrid Microwave Integrated Circuits

The term hybrid microwave integrated circuit refers to a single-substrate circuit "board" with passive components to which one or more active device chips are attached. We will restrict our attention to passive components and interconnects defined mostly by thin-film technology on suitable microwave substrates like alumina, sapphire, quartz, or BeO. Because of its high thermal conductivity, BeO is the preferred medium for power circuits. The microwave passive components may be in either lumped or distributed from, or a combination of the two. The active and even some passive components may be soldered, epoxy-bonded, or diffusion bonded. Interconnects to these attached active (and passive) components may be made by wire or ribbon bonding.

Note that if the attached components are in a suitable format such as beam-lead or solder bumped, a wire-bondless reflow-soldered attachment can be made.

These are characteristic features of HMICs:

(1) Active devices and passive circuits can be independently pretested before being committed to the circuit. This is particularly advantageous if the active devices are expensive or in short supply.

(2) In contrast to MMICs, active devices can be procured from various sources. Use of MMICs implies in-house capability.

(3) HMICs are more suited to medium- and higher-power applications than are MMICs, since insulating substrates with high thermal conductivity (e.g., BeO) can be used. It is also a common practice to attach active devices directly to a metal heat sink (usually the circuit frame) by suitable partitioning of the dielectric substrate. HMICs are the preferred choice in applications where a moderately large number of modules are required, in cases where high heat dissipation makes the use of GaAs as a substrate material marginal, and especially, although not exclusively, in applications with frequency coverage below X-band. Since, in hybrids, different materials and processing steps can be used for the circuits and the devices, each type can be optimized with regard to its specific requirements. The circuit substrates can be lower in cost and have better heat-dissipation properties and lower loss properties than the substrates used for device fabrication.

To achieve very small size, light weight, high reliability, and low cost, the ideal hybrid may take the following form: a single substrate (alumina, quartz, or BeO, as the application demands) carries all circuit components - lumped inductors, capacitors, resistors, etc. - in integrated form. Due to the use of lumped elements wherever possible, the substrate will be very small in size; it thus becomes suited for low-cost batch processing.

Active devices are flip-chip(solder bumped)-mounted to this substrate to form the complete operating module. GaAs FETs have the advantage, not found i

Si bipolars, of all contact points being accessible on the top surface. These contacts when equipped with plated bumps have the triple function of providing the mechanical support, the electrical connections, and the heat sink. Flip-chip bump mounting eliminates the commonly used wire bonds and thus reduces parasitics and the manual-labor content. The circuit substrates can have either wrap-around grounds or be equipped with via holes or special metal septa to provide the necessary isolation between stages and the low-inductance ground returns for the devices.

RCA is developing a new miniature hybrid technology called Miniature Beryllia Circuit (MBC) that combines the advantages of very small size, light weight, and the batch fabrication process of the monolithic approach with the flexibility of the hybrid approach (that permits the use of separately attached active devices) [147]. The MBC configuration uses a combination of distributed and lumped circuit elements defined on BeO to provide circuit substrates that combine high Q with excellent heat-dissipation properties. The BeO substrates - which are rough even when polished - are selectively glazed to provide smooth surfaces for the definition of high-quality factor lumped-element components including inductors, thin-film capacitors, and bridged interconnections. The active devices, GaAs power FETs in pellet form, are flip-chip mounted on unglazed sections of the BeO surface. A low-parasitic ground connection for the FET source contacts is provided by a metal septum that is fabricated as an integral part of the BeO substrate.

The major advantages of this technology are small-size, lightweight properties of the circuits combined with excellent heat dissipation and low loss. Wire bonds are completely eliminated. The use of batch fabrication processes for the circuits preserves many of the advantages of monolithic circuits, while maintaining flexibility and avoiding the need for employing expensive GaAs substrates for the passive circuits; this is all the more significant as passive circuits often are one to two orders of magnitude larger than the active devices. Since circuits and devices are fabricated on separate substrates, they can be better optimized for their intended functions. This is especially desirable for power applications, in which properties such as ease of mounting, good heat dissipation, and low circuit losses are of importance.

Figure III-37 shows the schematic of a 2-stage amplifier fabricated by means of this approach; figure III-38 is the photograph of a batch-fabricated chip; and figure III-39 shows some detail of the amplifier.

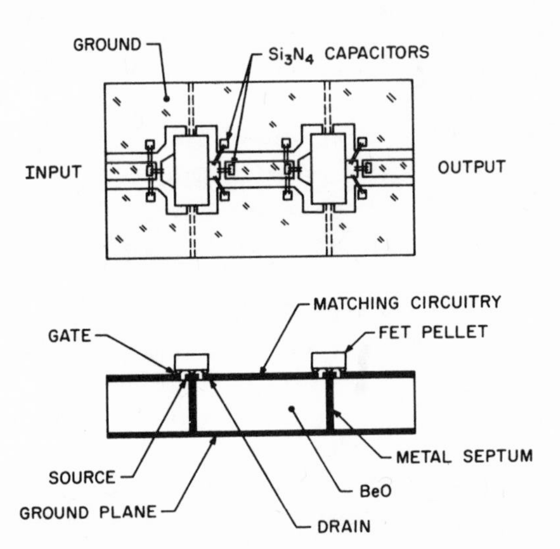

Figure III-37. Two-stage amplifier module.

Examples of modern HMICs are given by Belohoubek in his review paper [145].

2. Monolithic Microwave Integrated Circuits

The concept of MMICs goes back to 1964 and a government-funded program based on Si technology to develop an airborne transmit-receive module for a phased-array system. The results were not very impressive because of the inability of semi-insulating Si to maintain its resistivity through the many high-temperature diffusion processes involved. The development of the GaAs FET and the many improvements in GaAs materials and device technology have revitalized the subject. An excellent review of MMICs with many examples and detailed design considerations has been presented by Pucel [144].

The features of GaAs FETs and GaAs planar technology that make it attractive for MMICs have been mentioned throughout Section III.D. These features are summarized below.

(1) The superior frequency response, noise figure, linearity, and versatility of the planar GaAs FETs.

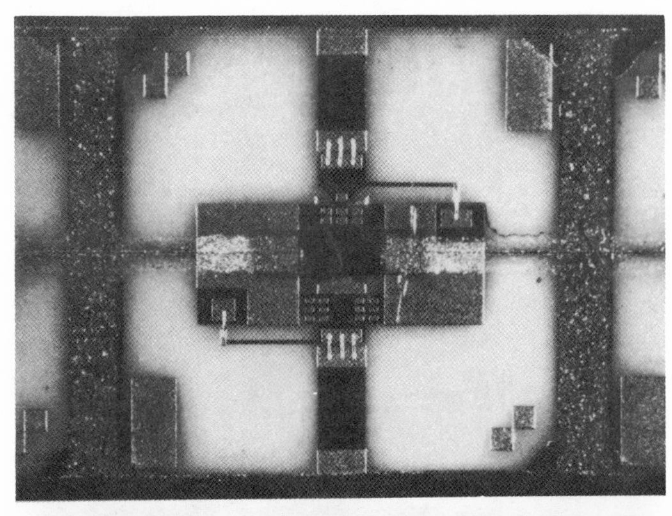

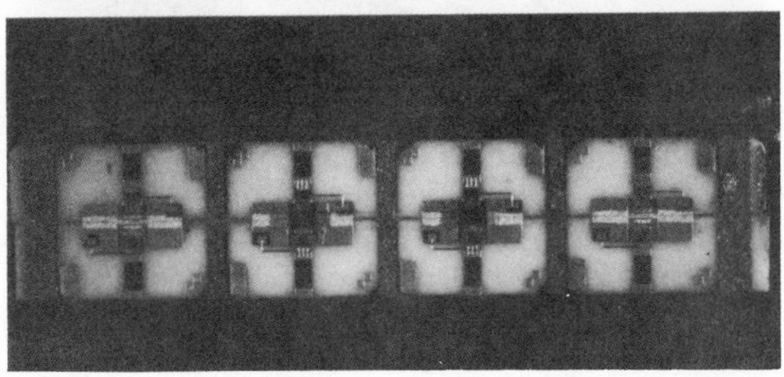

Figure III-38. Photograph of batch-fabricated chip containing two-stage amplifiers (see fig. III-37).

(2) The excellent dielectric properties of SI GaAs as a host for microwave transmission lines.

(3) Improvement in GaAs material technology, in particular, the availability of 2- to 3-in.-diam LEC-grown SI GaAs substrates suitable for direct ion implantation.

(4) Fabrication technology being developed for digital GaAs ICs is applicable to MMICs.

We will not discuss the design considerations and examples any further, since this material is available in the literature. Figure III-40 is an example of an MMIC phase shifter fabricated at RCA [148].

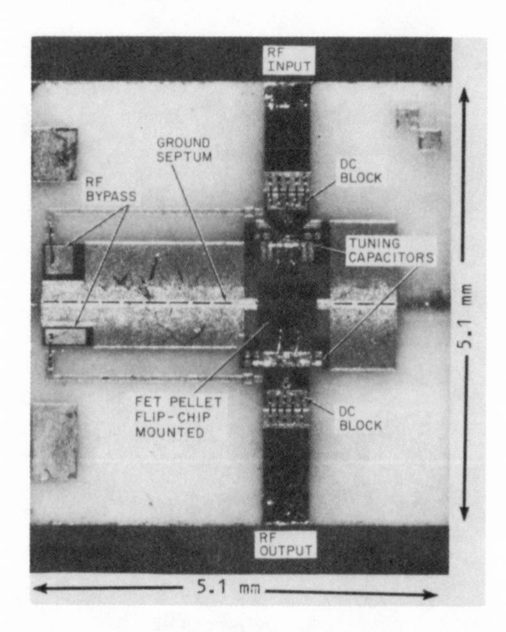

Figure III-39. Details of amplifier fabricated on beryllia substrate, shown in figure III-38.

E. EXPLORATORY DEVELOPMENT

This section describes the exploratory development efforts on III-V compound devices that have potential for impacting microwave and multigigabit-rate logic technology in the 1990 time frame. This exploratory development effort may be broadly classified under the following three categories:

(1) The search for materials with semiconducting properties superior to those of GaAs. The semiconductor properties of interest are high electron mobility, high peak electron velocity, and capability of growth on a suitable insulating substrate. $Ga_{0.47}In_{0.53}As$ lattice-matched to semi-insulating InP currently appears to be the most promising candidate.

(2) The exploitation of heterojunction structures, principally GaAs/GaAlAs structures. Examples are the development of heterojunction bipolar transistors (HBT) in which the wider-bandgap material serves as the emitter, and the use of single-period modulation-doped superlattices for FET fabrication (HEMT or

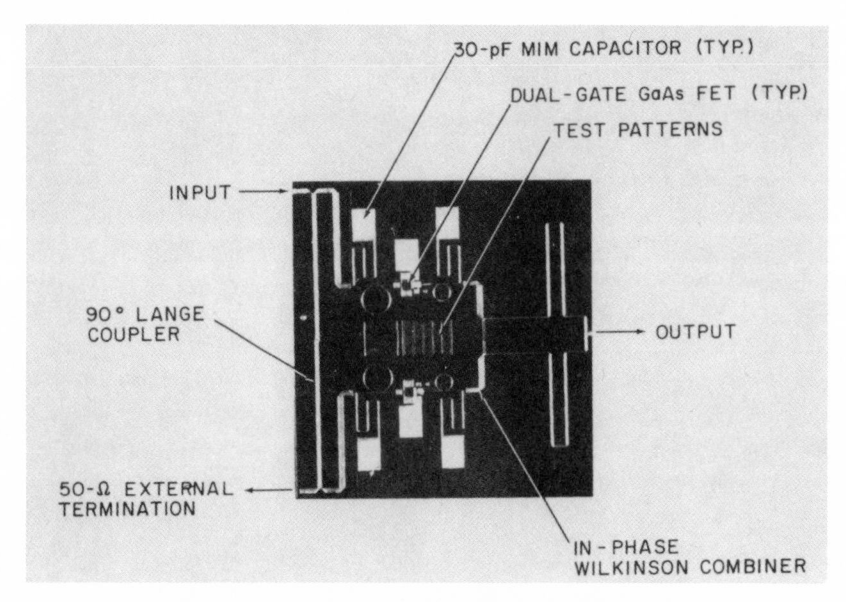

Figure III-40. Monolithic MIC phase shifter [148].

TEGFET). The potential of heterojunction bipolar structures has been elegantly treated by Kroemer [149].

(3) The exploitation of quasi-ballistic effects in submicrometer device structures. This includes, among others, conventional FETs with submicrometer gates, vertical FET geometries (permeable base transistor or PBT), and planar-doped barrier devices. These categories are briefly discussed below.

1. $Ga_{0.47}In_{0.53}As$ Devices

Theoretical calculations indicate that the $Ga_xIn_{1-x}As_yP_{1-y}$ alloy system, especially for compositions lattice-matched to InP (y = 2.2 x), may have potential for higher low-field mobility and peak drift velocity than has GaAs [134]. Experimental investigations show, however, that the electron mobility in this quaternary alloy system exceeds that in GaAs only near the ternary ($Ga_{0.47}In_{0.53}As$) limit [150].

The potential advantages of the $Ga_{0.47}In_{0.53}As/InP$ (referred to for simplicity as GaInAs/InP) alloy for high microwave frequency FETs are as follows:

(1) High low-field electron mobility. At a doping level of 10^{17} cm^{-3}, the electron mobility in GaInAs can approach 8000 cm^2 V^{-1} s^{-1}, compared to about 4500 cm^2 V^{-1} s^{-1} for typical GaAs layers [35,36]. This high mobility results in lower parasitic resistances and hence improved performance particularly for FETs operating at frequencies of 20 GHz and above.

(2) The peak electron drift velocity in GaInAs is 2.8×10^7 cm s^{-1} compared to 2×10^7 cm s^{-1} in GaAs [35]. Thus, for gate length limits set by technological constraints, GaInAs FETs have potential for superior frequency response.

(3) Calculations of the frequency response of submicrometer-gate-length FETs, including transient transport effects, show that to a first order, the transit time of electrons under the gate is inversely proportional to $\mu_o \Delta\varepsilon_{\Gamma L}$ where μ_o is the low field mobility and $\Delta\varepsilon_{\Gamma L}$ is the intervalley energy separation [151]. For GaInAs, $\mu_o \Delta\varepsilon_{\Gamma L}$ is 4760 cm^2 s^{-1} compared to 1520 cm^2 s^{-1} for GaAs, again indicating the potential of GaInAs for higher-frequency operation [151]. This figure of merit can be understood as follows: The low field mobility is the measure of velocity imparted to the carriers by the applied electric field. $\Delta\varepsilon_{\Gamma L}$ is the measure of threshold energy at which carriers are scattered to the lower mobility sub-band and velocity begins to saturate. $\Delta\varepsilon_{\Gamma L}$ is about 0.55 eV for GaInAs compared to 0.32 for GaAs [152]. <u>This combination of high mobility, low electron effective mass, and large ΓL separation is unknown in any other semiconductor whose bandgap is large enough to allow use at room temperature [152]</u>. Figure III-41 shows the computed cut-off frequency of GaInAs transistors compared to that of comparable GaAs and Si devices [151]. These calculations include transient (velocity overshoot) effects.

(4) As in the case of GaAs FETs, a semi-insulating substrate (Fe-doped InP) is available. Such a substrate is essential for realizing transistors for high microwave frequencies. Furthermore, since the bandgap of InP (1.34 eV) is greater than that of the ternary layer (0.72-0.75 eV), the heterojunction interface may assist in confining the electrons to the active layer. This is analogous to the use of GaAlAs buffer layers for GaAs FETs.

The potential disadvantage of GaInAs compared to GaAs for FET applications is that its bandgap is of the order of 0.72-0.75 eV. Typical metal-GaInAs Schottky-barrier heights are about 0.3 eV [153]. In order to circumvent

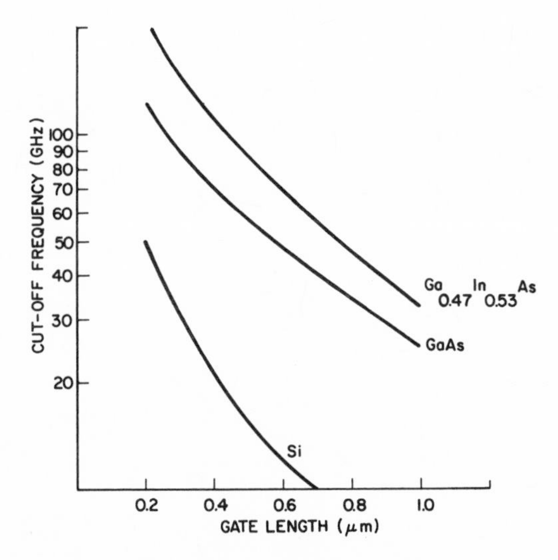

Figure III-41. Computed performance of FETs, including velocity overshoot.

problems due to this low Schottky-barrier height, various approaches including p^+n junction gates [154], heterojunction gates [155], and MIS gates [36], are being investigated at many laboratories.

Figure III-42 shows the room temperature electron mobility, μ(300), of a number of samples (grown at RCA) plotted as a function of the free-electron density [150]. The line shows the typical electron mobility in GaAs for a compensation factor, $(N_D + N_A)/(N_D - N_A)$, equal to 2. It is clear that this ternary layer has higher electron mobility than has GaAs of equivalent doping.

Very encouraging results have been obtained at RCA on GaInAs MISFETs fabricated by the use of CVD SiO_2 as the gate oxide [36]. Field-effect mobilities as high as 5200 $cm^2\ V^{-1}\ s^{-1}$ have been obtained on deep-depletion GaInAs MISFETs [36]. These devices have shown excellent depletion and enhancement characteristics, and MISFETs of 1.5-μm gate length have given 4-dB gain at 6 GHz with a power output of 100 mW and 30% power-added efficiency.

Initial investigations of the SiO_2/ternary interface have shown interface state densities of about 2.5×10^{12} to 2×10^{10} $cm^{-2}\ eV^{-1}$ with time constants of

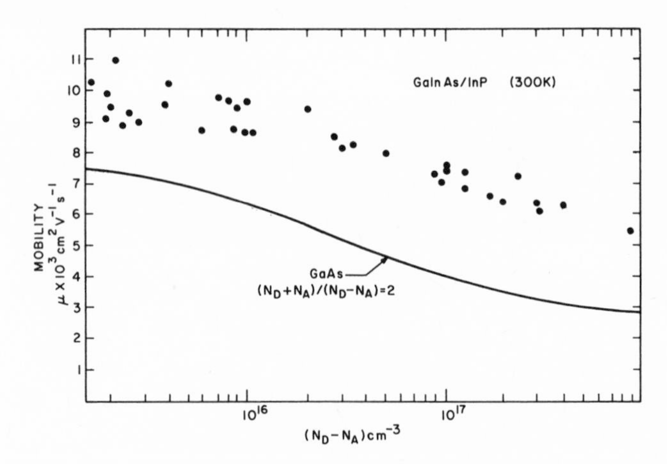

Figure III-42. Electron mobility of $Ga_{0.47}In_{0.53}As$ at 300 K as a function of carrier concentration [150].

1-100 μs. Effective mobilities of 5200 $cm^2\ V^{-1}\ s^{-1}$ have been measured, considerably greater than those reported for InP [156] and GaInAsP [157] quaternary MISFETs.

There are several potential advantages of this technology compared to that of GaAs MESFETs. Since the gate is a capacitor rather than a Schottky diode, the gate potential can swing equally positive as negative, as there is no forward conduction to limit the gate voltage. This enables devices to be operated at zero gate bias, simplifying the bias circuitry and leading to improved linearity. Also, if the improvement in mobility inherent in the material is realized, then devices with lower noise figures are possible [158].

From these results, it is evident that the Fermi level in this ternary layer is not pinned at midgap as in GaAs. Thus, inversion-mode enhancement devices are possible with this technology, making MIS-type logic a definite possibility. This would provide much simpler, lower-power gigabit-rate logic circuits than the circuits available to GaAs MESFET logic. In fact, several investigators [159,160] have already demonstrated that p-GaInAs can be inverted. RCA research has shown that inversion-mode mobilities as high as 600 $cm^2\ V^{-1}\ s^{-1}$ can be realized [161]. The potential exists for obtaining inversion-mode mobilities as high as 4000 $cm^2 V^{-1} s^{-1}$. The experience with Si MOS devices has shown that inversion-mode mobilities of 0.5-0.7 times the bulk mobilities are possible. Table III-23 indicates the potential of GaInAs vis à vis GaAs and Si for logic applications. Table III-24 summarizes the best results obtained to date with GaInAs microwave MISFETs [161]. Computer simulations also show that the minimum noise figure obtainable from GaInAs transistors may be 40-45% lower than that in equivalent GaAs devices [158].

Table III-23. COMPARISON OF Si, GaAs, AND GaInAs FOR LOGIC APPLICATIONS

Property	Si	GaAs	GaInAs/InP
Drift Mobility at 10^7 cm^{-3} ($cm^2 V^{-1} s^{-1}$)	1000	4500	8000
Peak Velocity (cm s^{-1})	$1x10^7$	$1.7x10^7$	$2.8x10^7$
Inversion-Mode MOS/MIS FETs	Yes	No	Yes
MOS/MIS Effective Mobility ($cm^2 V^{-1} s^{-1}$)	800	--	Potential for >4000
Delay (ps)	∿100 (MOS)	30 (d-MESFET) 100 (e-MESFET)	Potential for <50 (MIS)
Speed/Power Product	200 fJ (MOS)	1 pJ (d-MESFET) 5 fJ (e-MESFET)	Potential for <10 fJ (MIS)
Noise Immunity	High	Low	High
Power Supply Tolerance	High	Low	High
Packing Density	High	Low	High

Another potential impact of GaInAs MISFET technology is on integrated optics. GaInAsP lasers and LEDs span the 1- to 1.7-μm wavelength range where low loss fibers are available. The development of MISFETs will allow monolithic integration of optical and electronic functions. This subject is, however, beyond the scope of this study.

As discussed in Section III.A (Materials Technology), GaInAs can be grown by VPE, LPE, MOCVD, and MBE. Ion implantation into this material has also been demonstrated. Fe-doped SI InP substrates (1- to 2-in. diam) of reasonable quality, grown by the LEC process, are also becoming available. This therefore promises to be an exciting area of development with many potential applications.

Table III-24. RCA $Ga_{0.47}In_{0.53}As$ MICROWAVE MISFET PERFORMANCE [161]
(Gate Width = 600 μm; Drain-Source Voltage = 4 V)

Wafer	Gate Length (μm)	Frequency (GHz)	Gain (dB)	Power Out (mW)	Efficiency (%)
Q463	2.5	6	3	57	19.7
Q422 #1	2.0	6	2.5	105	17.3
Q691 B	1.5	6	4.6	116	30.1
Q691 B	1.5	11	3	90	22
Q691 B	1.5	12	3	60	11

2. Heterojunction Devices

The use of heterojunctions for high-frequency transistors was first put forward in the late 1940s. The relatively primitive semiconductor technology, however, prevented any significant experimental studies. As discussed in Section III.A, two new epitaxial growth technologies, viz. MBE and MOCVD, have emerged that offer the promise of routine growth of heterojunction structures. This development has rekindled interest in heterojunction microwave devices. As pointed out by Kroemer, the underlying central principle is to use the energy gap variations in addition to electric fields to control the forces on electrons and holes separately and independently of each other [149]. The two most important classes of heterostructures from our point of view are the single-period modulation-doped superlattice and the heterojunction bipolar transistor (HBT) with a wide-bandgap emitter. The former technology promises to have its major impact on multigigabit-rate logic; in particular, on logic circuits operating at 77 K with speed and power comparable to Josephson junction devices that operate at 4-5 K [162]. HBTs offer the potential of microwave transistors with maximum oscillation frequencies of 100 GHz and switching speeds of the order of 10 ps at 300 K [162].

a. Single-Period Modulation-Doped Structures - Super-Mobility Devices

The electron mobility in 10^{17} cm^{-3} doped GaAs homostructures does not increase too rapidly as the temperature is lowered to 77 K, since the mobility is limited by ionized impurity scattering. It was pointed out by Dingle et al. [163] that in a heterostructure like $GaAs/Ga_{1-x}Al_xAs$, the donor atoms can be put in the wider-bandgap $Ga_{1-x}Al_xAs$ layer adjacent to the nominally undoped

GaAs layer. Free electrons spill over from $Ga_{1-x}Al_xAs$ to the GaAs layer; there they are constrained by the conduction-band discontinuity. Since there are (ideally) very few ionized donors in the GaAs layer, mobilities corresponding to typical 10^{15}-cm^{-3} or lower-doped GaAs should be obtained even though there is a high sheet electron density ($\sim 10^{11}$-10^{12} cm^{-2}). For example, a high-quality, uniformly doped (10^{17} cm^{-3}) GaAs FET channel layer has a Hall mobility of 4500-5000 cm^2 V^{-1} s^{-1} at 300 K. In contrast, experimental mobilities as high as 8400, 100,000, and 210,000 cm^2 V^{-1} s^{-1} at 300, 77, and 10 K, respectively, have been obtained in a GaAs/GaAlAs heterostructure with comparable charge density [164]. FETs built with single-period GaAs/GaAlAs heterostructures have been variously called high-electron mobility transistors (HEMTs) and two-dimensional electron gas field-effect transistors (TEGFETs). With simple logic circuits based on these devices, propagation delays of 19 ps at 300 K and of 16 ps at 77 K have already been reported [162] .

Figure III-43 is a schematic representation of the band structure of a single-period modulation-doped GaAs/GaAlAs structure. At the interface, the band discontinuity occurs mainly at the conduction band edges. For example, the conduction band edge in $Al_{0.7}Ga_{0.3}As$ is roughly 300 MeV higher than in GaAs. Shallow donors incorporated in GaAlAs transfer to GaAs where they can have lower energy [163], thus creating a strong internal electric field. This leads to severe band-banding at the interfaces as shown in figure III-43, creating an almost triangular potential well wherein electrons can be trapped. The trapped electrons behave dynamically like a two-dimensional electron gas (2DEG), as demonstrated by magnetophonon resonance [165] and Shubnikov-deHaas measurements [166]. If the GaAs doping is low ($\sim 10^{14}$-10^{15} cm^{-3}), these electrons are not subjected to impurity scattering, and mobility is limited by the properties of low-doped GaAs, interface roughness, and Coulomb interaction with donors across the interface.

Figure III-44 shows schematic cross sections of two types of modulation-doped heterostructures currently being studied. The semiconductor structure shown in figure III-44(a), with the ternary on top, usually leads to higher mobilities. Figure III-45 shows the schematic of a HEMT (or TEGFET).

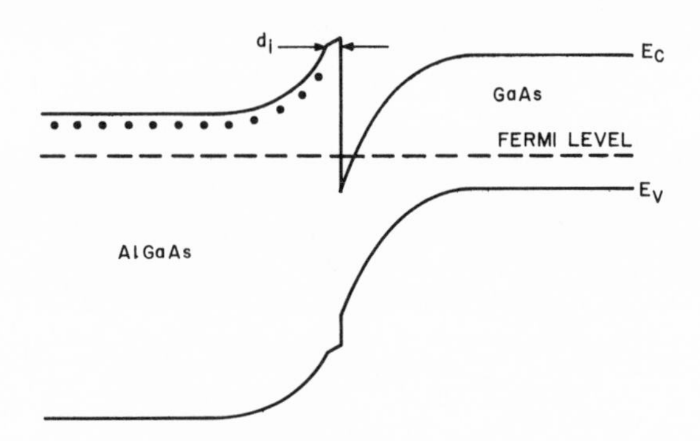

Figure III-43. A schematic representation of the band structure of a modulation-doped GaAs/AlGaAs heterojunction.

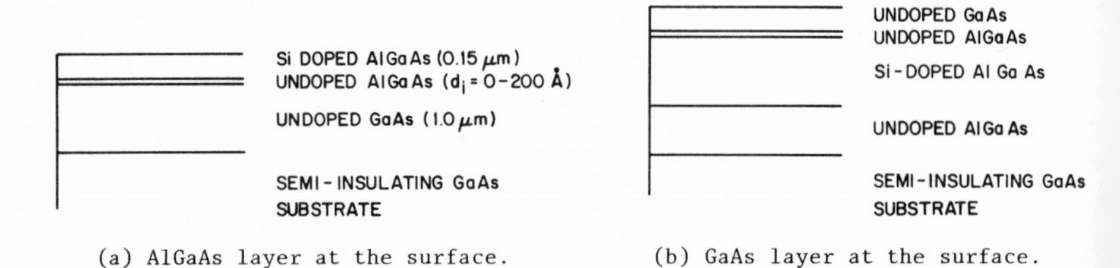

(a) AlGaAs layer at the surface. (b) GaAs layer at the surface.

Figure III-44. A schematic cross section of modulation-doped heterostructures.

The impact of the development of HEMT (or TEGFET) technology is not quite clear. The potential areas of application are as follows:

(1) 300 K low-noise and power FETs for frequencies greater than 20 GHz. Even the moderate increase in 300 K mobility may provide the marginal increase in performance to make these devices viable.

(2) 77 K low-noise FETs for use in receivers. The improved noise figure due to higher mobility may make these FETs very competitive to cooled paramps.

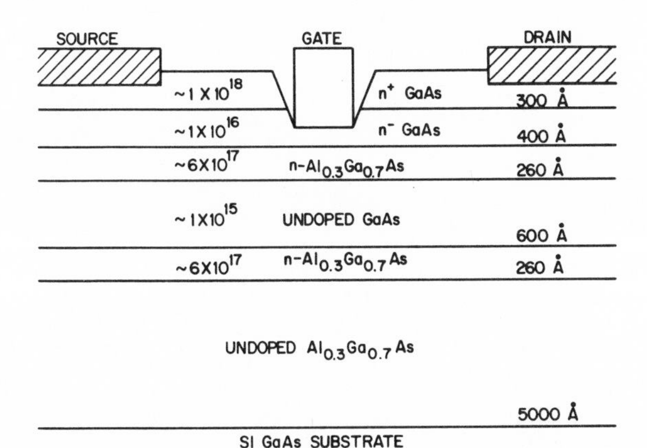

Figure III-45. HEMT (or TEGFET) structure.

(3) Multigigabit-rate logic applications at 300 and 77 K. At 77 K, the HEMT may seriously challenge Josephson junction technology.

b. Heterojunction Bipolar Transistors (HBTs)

An excellent review of HBTs and their impact on microwave and multigigabit-rate logic technology has recently been presented by Kroemer [149]. As indicated earlier, the principle of wide-bandgap emitters is well known - only now the technology for exploiting this effect is at hand. The maximum current gain β_{max} of an HBT is given by the equation [149]

$$\beta_{max} = \frac{N_D}{N_A} \frac{v_{nb}}{v_{pe}} \exp(\Delta E_g/kT) \tag{8}$$

where N_D and N_A are the emitter and base doping in an n-p-n transistor, v_{nb} and v_{pe} are the mean carrier velocities in the base and emitter region (including drift and diffusion), and ΔE_g represents the amount by which the emitter bandgap exceeds the base bandgap. In a homojunction bipolar transistor (BJT), ΔE_g is zero; the exponential factor thus represents the effect of the wide-bandgap emitter. When ΔE_g is larger than kT, the exponential term can be significant. For example, kT = 0.026 eV at room temperature, and a ΔE_g of 0.2 eV thus represents an exponential factor of 3000. In a BJT the emitter doping is about two orders of magnitude higher than the base doping to obtain good injection efficiency. In an HBT, this ratio can be inverted, allowing base doping of 10^{19} cm^{-3} and yet maintaining a high β. This will result in a significant lowering of base resistance R_b. The current cut-off frequency (f_T) of a bipolar transistor is given by

$$f_T = 1/(2\pi R_b C_c) \tag{9}$$

where C_c is the collective capacitance. Decreasing the base resistance increases the current cut-off frequency and improves frequency response. The various design tradeoffs in an HBT are discussed by Kroemer [149]. Kroemer also points out that to fully exploit the decreased base resistance, suitable device structures must be designed. It is not enough to just replace the emitter with a wider-bandgap material. He indicates that a heterojunction base-collector interface may offer some significant advantages [149].

The potential advantages of the HBT compared to field-effect transistors are as follows:

(1) The speed (frequency response) determining part of the current path is perpendicular to the semiconductor surface in a bipolar transistor. To the first order, speed is then determined by layer thickness (rather than lithography, as for an FET). Since vertical thicknesses can be made much smaller than horizontal lithography dimensions, HBTs have inherently higher speed potential.

(2) The smaller critical current paths may make it easier to exploit ballistic effects. Furthermore, the emitter junction barrier represents a "ballistic ramp" allowing high-velocity carrier injection.

(3) For multigigabit-rate logic applications, particularly for LSI and VLSI circuits, threshold voltage uniformity and reproducibility is very critical. For HBTs (and BJTs), the threshold voltage depends on base bandgap and, weakly (logarithmically), on base doping. On the other hand, in FETs the threshold voltage depends on the pinchoff voltage, which in turn is proportional to Nd^2 and is technologically much harder to control.

The potential advantages of the HBT must be traded off against the somewhat simpler FET technology.

3. Quasi-Ballistic Devices

With the improvement in semiconductor material, purity, and the development of fabrication technologies that permit micrometer and submicrometer critical dimensions, hot electron effects are becoming more and more important in semiconductor devices. These new developments are hampered by a lack of understanding of the physics of nonequilibrium electron transport on a scale intermediate to the true atomic (~10 Å) and the bulk solid-state (<1 μm) regimes. This is a subject of some controversy and debate, and the August 1981 issue of the IEEE's Transactions on Electron Devices was wholly devoted to this topic [167].

Most of the questions that arise about transport in submicrometer devices are due to the extremely fast time scales involved. For example, electrons traveling at 10^7 cm s^{-1} will traverse a 0.1-μm channel in 10^{-12} s, a time scale of the order of the momentum, energy, and charge relaxation times. On such a time scale, electrons encountering a high-field region do not have adequate time to establish any type of equilibrium distribution and may "overshoot" the peak velocity computed on the basis of conventional theory. This point was first made by Ruch in 1972 [92], and later the occurrence of velocity overshoot in GaAs was established experimentally [168].

There have been several publications [169,170] suggesting the possibility of purely ballistic or inertial (collisionless) transport in devices with submicrometer features. This assertion has been the subject of some controversy [167] that exceeds the scope of this review. However, we will summarize some tentative conclusions that may impact the design and performance of high-speed devices, as follows:

(1) The investigation of nonequilibrium effects shows that velocity overshoot over the "peak" average drift velocity calculated by conventional theory does occur [168]. The speed of semiconductor devices can thus be increased by exploiting this phenomenon. In GaAs, overshoot effects probably begin at channel lengths just shorter than a micrometer.

(2) Purely ballistic operation (nearly collisionfree transport) occurs in GaAs only for voltages in the 0.04-V regime [171]. Other semiconductors such as GaInAs or InAs may have more attractive properties insofar as purely ballistic devices are concerned.

(3) Suitably tailored electric fields may enhance velocity overshoot effects. A δ-function-like field at a source contact that rapidly accelerates

the carriers over a small distance and then injects them into a sustaining field of suitable magnitude is one strategy to obtain very high carrier velocity (planar doped barrier, PDB) [171].

(4) In devices with submicrometer critical dimensions, diffusion effects, particularly in proximity to contacts, will have a significant effect and must be factored into any analysis.

(5) Quantum-mechanical effects will have to be included. Semiclassical approaches based on the Boltzmann transport equation may not be adequate [167].

a. Effect on Classical Planar FETs

The simplest class of devices in which nonequilibrium transit phenomena (e.g., velocity overshoot) become important are FETs with micrometer or submicrometer channel lengths. This was first pointed out by Ruch [92], whose one-dimensional analysis showed that overshoot effects become important for channel lengths of <1 µm and 0.1 µm for GaAs and Si devices, respectively. A recent comparison of FET performance by Cappy et al. [151] shows that in GaInAs and InAs overshoot effects may be more pronounced than in GaAs. Figure III-46, taken from reference 152, illustrates this point. Table III-25 shows the computations of Curtice for GaAs MESFETs as the gate length is decreased [173]. These calculations include both two-dimensional and velocity-overshoot effects obtained with a phenomenological model [174]. GaAs MESFETs with 0.7-µm gate length, fabricated at RCA, show f_T in the 36- to 40-GHz regime, values in reasonable agreement with Curtice's calculations.

b. Permeable-Base Transistor (or Vertical GaAs FETs)

The previous section described the effect of velocity overshoot on conventional planar FETs. The gate length in these structures is determined by lithographic constraints; <0.5-µm gate lengths will require deep-UV lithography, x-ray lithography, or direct-write electron-beam lithography. Researchers at MIT have proposed a structure called the permeable-base transistor (PBT) [175], wherein the critical gate length is determined by the metal thickness, which, in principle, can be controlled to 0.05 µm or even lower. Practical limits such as metallization resistance rather than technological constraints will set the lower limit. Although this structure is fairly complex, its proponents predict that devices with f_T exceeding 100 GHz are possible [175]. Notwithstanding the technological complexity, this structure is perhaps the most promising to fully exploit velocity overshoot.

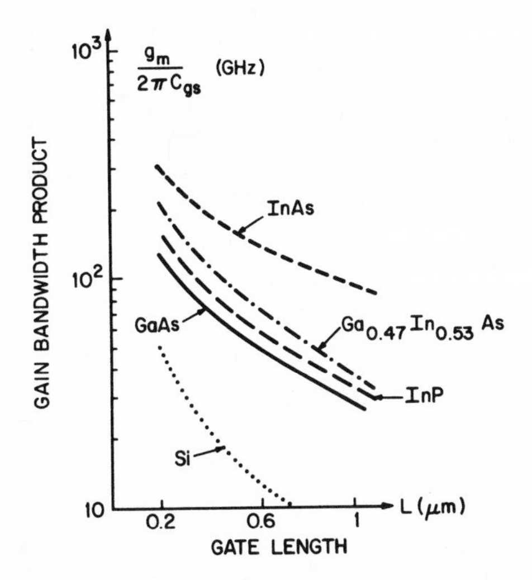

Figure III-46. Gain-bandwidth product vs gate length [151]. © 1980 IEEE.

Table III-25. EFFECT OF GATE LENGTH ON GaAs MESFET CUT-OFF FREQUENCY [173] (Velocity Overshoot Included)

$N = 1.0 \times 10^{17}$ cm^{-3} $V_{DS} = 3.0$

$d = 0.16$ μm $V_{GS} = -1.25$ V

$W = 200$ μm (total)

Gate Length (μm)	f_T (GHz)	F_{max} (GHz)
1.5	16.6	83.5
1.0	23.8	106
0.76	31.2	108
0.5	42.1	132
0.36	50.9	158
0.36*	65.2	207

*d = 0.14 μm.

A three-dimensional drawing of the PBT showing a cutaway region is given in figure III-47. The unique feature is the thin tungsten grating, which consists of a parallel array of tungsten stripes connected together and embedded in a single-crystal semiconductor. The tungsten grating separates the emitter layer from the collector layer and forms the base of the transistor. Because the tungsten is completely surrounded by the semiconductor and forms a Schottky barrier with it, the voltage on the metal can be used to control the current flowing between the stripes from the emitter layer into the collector layer. Proton bombardment can be used to transform the gallium arsenide into semi-insulating material, and for the device in figure III-47, the proton-damaged regions minimize parasitic capacitance and isolate neighboring devices.

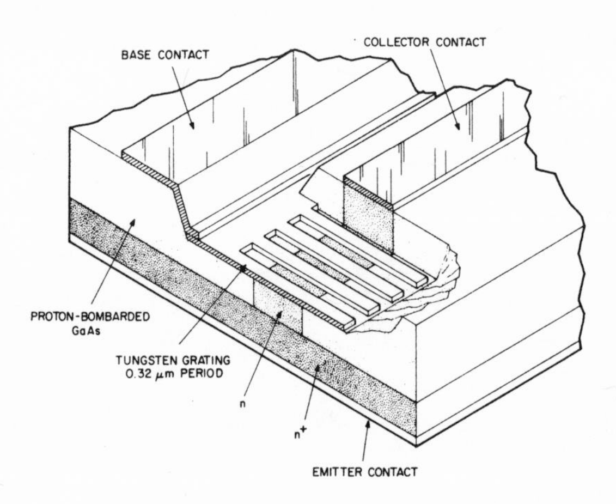

Figure III-47. Three-dimensional drawing of a GaAs PBT. A grating made from a 300-Å-thick film of tungsten is embedded in the single crystal. The electrons flow from the emitter layer to the collector layer through the openings in the grating [175]. © 1980 IEEE.

In initial devices reported, the grating consisted of 1600-Å (0.16 μm) stripes and spaces patterned in a 300-Å(0.03 μm)-thick tungsten layer. The carrier concentration in both the emitter and collector layers is approximately

$3x10^{16}$ cm^{-3}, with thicknesses of 0.3 and 2.0 μm, respectively. Although the PBT has so far been made only in gallium arsenide, one should also be able to fabricate the device in silicon. The results of the simulations for this device indicate that the switching time delay in an inverter gate with a fan-out of one could be as low as 2 ps. Although an inverter gate has not yet been built, small signal time delays of 4.3 ps have been calculated from microwave network analyzer measurements of present devices.

It should be noted that while submicrometer gate length can be controlled easily by metal thickness, submicrometer lithography is still required. Epitaxial overgrowth over metal is a complex technology that requires considerable development. Finally, for logic applications, good threshold-voltage uniformity in PBTs will be difficult to obtain since threshold voltage is highly geometry sensitive. At present, we believe that the PBT may be more suited for microwave rather than logic applications. The PBT is discussed in reference 175 for microwave and in reference 176 for logic applications.

c. Planar-Doped Barrier Transistors

Attempts to use an inhomogeneous field-distribution to rapidly accelerate electrons over very short distances and inject them through a thin base region are based on the planar-doped barrier transistor concept [172]. By this strategy one can, in principle, prevent significant cooling of the injected "hot" carriers.

The planar-doped barrier (PDB) structure uses a plane of p-type dopant positioned within an undoped region that is bounded by two n^+contacts, as shown in figure III-48. The p doping is much higher than the n doping of the background-doped regions. The p region will, within certain constraints, be fully depleted and, at zero bias, narrow space-charge regions will be induced in the two n^+ regions to satisfy the condition for charge neutrality. The proportions of this space charge in regions A and B will depend on the position of the p plane within the undoped region. Thus, the charge profile will be that indicated in figure III-48(c), and a solution of Poisson's equation will yield the field and energy band diagrams shown in figures III-48(d) and (e). Under positive bias (layer A positive), and when the plane is assumed to be much closer to A than to B, barrier ϕ_{B1} will increase slightly and barrier ϕ_{B2} will be reduced by an amount nearly equal to the bias voltage, giving rise to forward conduction by electrons being emitted from layer B over the barrier to the surface contact. With negative bias, ϕ_{B1} is reduced only slightly and ϕ_{B2} increases almost by the

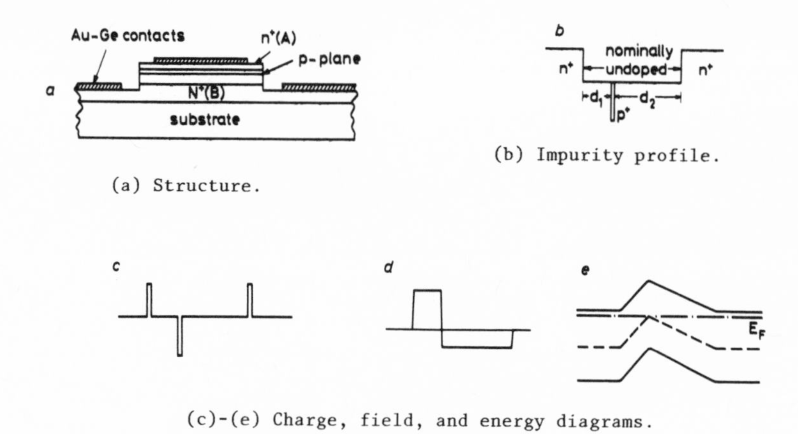

(a) Structure.

(b) Impurity profile.

(c)-(e) Charge, field, and energy diagrams.

Figure III-48. Planar-doped barrier device [177]. Courtesy IEE Professional Publications.

amount of the applied voltage. Reverse current flow will therefore occur at a much higher voltage, this time by the emission of electrons from layer A over the barrier into the substrate. The I-V characteristic can therefore be made symmetrical, forward with positive or with negative bias, by varying the position of the p plane within the undoped region.

The proposed transistor consists of two PDBs joined together by a common base region. This device uses a forward-biased emitter barrier to accelerate electrons to 0.3 eV over a distance of a few hundred angstroms, and injects these electrons into a thin base region with a thickness of up to a few thousand angstroms. Most of these hot electrons are not significantly cooled during their transit through the base, and thus surmount the reverse-biased collector barrier. Since the injected electrons are accelerated over very short time and distance scales, it is predicted that they will have very high velocities, approaching the crystal limited group velocity of $\sim 1\times10^{8}$ cm s^{-1} in GaAs. As a result of the high electron velocities and majority-carrier conduction, the PDB transistor promises to be a very high speed device.

Prototype transistor structures have been fabricated in GaAs layers grown by MBE [172]. Preliminary I-V measurements have demonstrated dc current gains of about 20. The design criteria and tradeoffs, and the fabrication technology required for the development of these transistors, are not yet well established.

d. Discussion

The exploitation of ballistic (or quasi-ballistic) effects in semiconductor devices is an exciting new field that is still controversial. While it is clear that velocity-overshoot effects in conventional devices such as submicrometer-gate FETs will lead to some improvement in speed, the potential of novel devices is hard to evaluate with any degree of confidence. The analysis and modeling of these devices is fairly complex and requires further study.

The state of the art of III-V-compound technology and the current direction of research has been elegantly summarized by Kroemer [149]. The following two paragraphs are taken from his paper:

> "We have witnessed, since about 1964, a steady growth in III/V compound semiconductor devices, principally GaAs devices. The driving force behind this development has been the high performance of such devices, not attainable with mainstream Si devices. If we ignore once again lasers and other optoelectronic devices, and restrict ourselves to purely electronic amplifying and switching devices, high performance has been largely synonymous with high speed, made possible by the high electron mobility of GaAs, and by the availability of semi-insulating GaAs as a substrate. However, not even the most ardent advocate of GaAs ever claimed that GaAs was used because it had an attractive technology. We used GaAs despite its technology, not because of it, and the threat was never far away that Si devices, with their much simpler and more highly-developed technology, would catch up with GaAs performance, the fundamental advantages of GaAs notwithstanding.
>
> "It is exactly this imbalance between fundamental promise and technological weakness that is being removed by the new epitaxial technologies.* If the technological scenario postulated in Section III of this paper is even remotely correct, it means nothing less but that the great future strength of III/V compounds lies precisely in their new technology, which permits an unprecedented complexity and diversity in epitaxial structures, going far beyond anything available in Si technology! This new technological strength is thus emerging as more important than the older fundamental

*MBE and MOCVD.

strengths of high mobilities and semi-insulating substrates. It is a remarkable reversal of priorities indeed."

F. REFERENCES

1. Rees, G. J., Editor: Semi-insulating III-V Materials. Shiva Publications, England, 1980.
2. Fairman, R. D., et al: Growth of High-Purity Semi-Insulating Bulk GaAs for Integrated Circuit Applications. IEEE Trans. Electron Devices, vol. ED-28, no. 2, Feb. 1981, pp. 135-139.
3. Hobgood, H. M., et al.: High-Purity Semi-Insulating GaAs Material for Monolithic Microwave Integrated Circuits. IEEE Trans. Electron Devices, vol. ED-28, no. 2, Feb. 1981, pp. 140-149.
4. Magee, T. J., et al.: Gettering of Cr in GaAs by Back-Surface Mechanical Damage. Phys. Stat. Sol. (a), vol. 55, 1979, p. 169.
5. Rossel, P., et al.: Effect of Substrate on Properties of GaAs FETs (in French). Rev. Phys. Appl., vol. 13, 1978, p. 503.
6. Zylbersztejn, A., et al.: Hole-Traps and Their Effects in GaAs MESFETs. Inst. Phys. Conf. Series, no. 45, 1979, pp. 315-325.
7. Upadhyayula, L. C., RCA Laboratories: Private Communication.
8. Crystal Specialties Inc., Monrovia, Calif.: Private Communication.
9. White, A. M., Whither Cr in GaAs?; in Rees, G. J., Editor: Semi-Insulating III-V Materials. Shiva Publications, England, 1980, pp. 3-12.
10. Blakemore, J. S.: Intrinsic Density $n_i(T)$ in GaAs. J. Appl. Phys., vol. 53, no. 1, Jan. 1982, pp. 520-53.
11. Martin, G. M.: Key Electrical Parameters in Semi-Insulating Materials: The Methods to Determine Them in GaAs; in Rees, G. J., Editor: Semi-Insulating III-V Materials. Shiva Publications, England, 1980, pp. 13-28.
12. Blakemore, J. S.: Modelling of a Multivalent Impurity, Such as GaAs; Cr. In Rees, G. J., Editor: Semi-Insulating III-V Materials. Shiva Publications, England, 1980, pp. 29-40.
13. Au Coin, T. R., et al.: Liquid Encapsulated Compounding and Czochralski Growth of SI GaAs. Solid-State Technol., vol. 22, no. 1, Jan. 1979, pp. 59-62.
14. Brice, J. C., et al.: Mass Spectrometric Studies of Impurities in GaAs. J. Mat. Sci., vol. 2, 1967, p. 131.

15. Brice, J. C.: The Effect of As Pressure on Crystal Efficiency for Injection Luminescence in GaAs. Solid-State Electron., vol. 10, 1967, p. 335.

16. Metz, E. P., et al.: A Technique for Pulling Single Crystals of Volatile Materials. J. Appl. Phys., vol. 33, 1962, p. 2016.

17. Mullin, J. B., et. al.: Liquid Encapsulation Techniques: The Use of An Inert Liquid in Suppression of Dissociation During Melt Growth of InAs and GaAs Crystals. J. Phys. Chem. Solids, vol. 26, 1965, 782.

18. Jordan, A. S., et al.: A Thermoelectric Analysis of Dislocation Generation in Pulled GaAs Crystals. Bell Syst. Tech. J., vol. 59, no. 4, 1980, p. 593.

19. Stinermann, A., and Zimmerli, U.: Dislocation-free GaAs Single Crystals. Proc. Inst. Cryst. Growth Conf., Boston, Mass., 1966, p. 81.

20. Asbeck, P., et al.: Effects of Cr Redistribution on Device Characteristics in Ion Implanted GaAs ICs Fabricated with SI GaAs. IEEE Trans. Electron Devices, vol. ED-26, no. 11, 1974, p. 1853.

21. Magee, T. J., et al.: Front Surface Control of Cr Redistribution and Formation of Stable Cr Depletion Channels in GaAs. Workshop on Process Technology for Direct Ion Implantation into SI GaAs, Santa Cruz, Calif., Aug. 1980.

22. Rumsby, D., et al.: Growth and Properties of Large SI Crystals of InP; Semi-Insulating III-V Materials. Shiva Publications, England, 1980, pp. 59-67.

23. Richman, D.: Dissociation Pressures of GaAs, GaP, and InP and Nature of III-V Melts. J. Phys. Chem. Solids, vol. 24, 1963, p. 1134.

24. Bachmann, K. J., et al.: Liquid Encapsulated Czochralski Pulling of InP Crystals. J. Electron. Mat., vol. 4, 1975, p. 389.

25. Antypas, G. A.: Preparation of High-Purity Bulk InP. Inst. Phys. Conf. Series, no. 33b, 1977, pp. 55-59.

26. Straughan, B. W., et al.: Eutectic Formation in Cr-Doped InP. J. Cryst. Growth, vol. 21, 1974, p. 117.

27. Mizuno, O., and Watanabe, H.: Semi-Insulating Properties of Fe-Doped InP. Electron. Lett., vol. 11, 1975, p. 148.

28. Williams, F. V., and Ruehrwein, R. A.: J. Electrochem. Soc., vol. 108, 1961, p. 117c.

29. Nelson, H.: Epitaxial Growth from the Liquid State and Its Application to the Fabrication of Tunnel and Laser Diodes. RCA Rev., vol. 24, 1963, p. 603.
30. Knight, J. R., et al.: Preparation of High-Purity GaAs by Vapor Phase Epitaxial Growth. Solid-State Electron., vol. 8, 1965, p. 178.
31. Davey, J. E., and Pankey, T.: Epitaxial GaAs Films Deposited by Vacuum Evaporation. J. Appl. Phys., vol. 39, 1968, p. 1941.
32. Manasevit, H. M., and Simpson, W. I.: Use of Metal Organics in the Preparation of Semiconductor Materials. J. Electrochem. Soc., vol. 120, 1973, p. 135.
33. Olsen, G. H., and Zamerowski, T. J.: Vapor-Phase Growth of (In,Ga)(As,P) Quaternary Alloys. IEEE Trans. Quantum Electron., vol. QE-17, 1981, p. 128.
34. Kressel, H., and Butler, J. K.: Semiconductor Lasers and Heterojunction LEDs. Academic Press, New York, 1977.
35. Littlejohn, M. A., Hauser, J. R., and Glisson, T. H.: Velocity-Field Characteristics of $Ga_xIn_{1-x}As_yP_{1-y}$ Quaternary Alloys. Appl. Phys. Lett., vol. 30, 1977, p. 242.
36. Gardner, P. D., Narayan, S. Y., Colvin, S., and Yun, Y. H.: $Ga_{0.47}In_{0.53}$ Metal Insulator Field-Effect Transistors (MISFETs) for Microwave Frequenc Applications. RCA Rev., vol. 42, Dec. 1981, p. 542.
37. Crystal Specialties Inc., Monrovia, Calif.: Product Sheet.
38. Cambridge Instruments Inc., Monsey, N.Y. (U.S. Rep. for Metals Research Ltd., UK): Product Sheet.
39. Keller, S. P., Editor: Handbook on Semiconductors, Vol. III. North-Holland, Amsterdam, 1980.
40. Ettenberg, M., et al.: Vapor Growth and Properties of AlAs. J. Electrochem. Soc., vol. 118, 1971, p. 1355.
41. Stringfellow, G. B.: OMVPE Growth of $Al_xGa_{1-x}As$. J. Cryst. Growth, vol. 55, no. 1, 1981, pp. 42-52.
42. Shaw, D. W.: Epitaxial GaAs Kinetic Studies. J. Electrochem. Soc., vol. 117, no. 5, 1970, pp. 683-687.
43. Hollan, L., et al.: Influence of Growth Parameters in GaAs Vapor Phase Epitaxy. J. Electrochem. Soc., vol. 124, no. 1, 1977, pp. 135-139.
44. DiLorenzo, J. V.: Vapor Growth of GaAs. J. Crystal Growth, vol. 17, 1972, pp. 189-206.

45. Nozaki, T., et al.: Multi-Layer Epitaxial Technology for Schottky Barrier GaAs FET. Inst. Phys. Conf. Series, no. 24, 1975, pp. 46-54.
46. Cox, H. M., and DiLorenzo, J. V.: Characteristics of an $AsCl_3/H_2/Ga$ Two-Bubbler GaAs CVD System for MESFET Applications. Inst. of Phys. Conf. Series, no. 33b, 1977, pp. 11-22.
47. Weiner, M. E.: Si Contamination in Open Flow Quartz Systems for Growth of GaAs and GaP. J. Electrochem. Soc., vol. 119, no. 4, 1972, pp. 496-504.
48. DiLorenzo, J. V., and Moore, Jr., G. E.: Effects of $AsCl_3$ Mole Fraction on Incorporation of Ge, Si, Se and S into VPE layers of GaAs. J. Electrochem. Soc., vol. 118, no. 11, 1972, pp. 1823-1830.
49. DiLorenzo, J. V.: Vapor Growth of Epitaxial GaAs: Summary of Parameters Which Influence the Purity and Morphology of Epitaxial Layers. J. Cryst. Growth, vol. 17, 1972, pp. 189-206.
50. DiLorenzo, J. V., and Machala, A. E.: Orientation Effects on Electrical Properties of High Purity Epitaxial GaAs. J. Electrochem. Soc., vol. 118, no. 9, 1972, pp. 1516-1517.
51. Mizuno, O., et al.: Epitaxial Growth of Semi-Insulating GaAs, Jpn. J. Appl. Phys., vol. 19, 1971, p. 208.
52. Jolly, S. T., et al.: Epitaxial Growth of SI GaAs. Annual Report, Contract No. N00014-77-C-0542, July 1978.
53. Cox, H. M., and DiLorenzo, J. V.: Review of Techniques for Epitaxial Growth of High Resistivity GaAs; in Rees, G. J., Editor: Semi-Insulating III-V Materials. Shiva Publications, 1980, England.
54. Steele, S. R., et al.: Growth and Characterization of p-type GaAs. Inst. Phys. Conf. Series, no. 45, London, 1979, pp. 45-51.
55. Komeno, J., et al.: Ultra-High Uniform GaAs Layers by VPE. Inst. Phys. Conf. Series, no. 56, London 1981, pp. 9-18.
56. Enstrom, R. E., et al.: Vapor-Phase Growth Technique for Several III-V Compound Semiconductors. Final Report, Contract No. NAS 12-538, April 1971.
57. Olsen, G. H., et al.: Crystal Growth and Properties of Binary, Ternary, and Quaternary (In,Ga)(As,P) Alloy Grown by the Hydride Vapor Phase Epitaxy Techniques; in B. R. Pamplin, Editor: Progress in Crystal Growth and Characterization, Vol. II. London, England, Pergamon, 1980.

58. Olsen, G. H., et al.: Vapor-Phase Growth of (In,Ga)(As,P) Quaternary Alloys. IEEE J. Quantum Electron., vol. QE-17, 1981, p. 128.

59. Ban, V. S.: Mass Spectrometric and Thermodynamic Studies of the CVD of Some III-V Compounds. J. Cryst. Growth, vol. 17, 1972, pp. 19-23.

60. Bass, S. J.: Device Quality Epitaxial GaAs Grown by the Metal Alkyl Technique. J. Cryst. Growth, vol. 31, 1975, pp. 172-178.

61. J. Cryst. Growth, vol. 55, no. 1, Oct. 1981. (Special Issue on metal-organic vapor-phase epitaxy.)

62. Dapkus, P. D., et al.: High Purity GaAs Prepared by Trimethyl Ga and AsH_3. J. Cryst. Growth, vol. 55, no. 1, 1981, pp. 10-23.

63. Nakanisi, T., et al.: Growth of High-Purity GaAs Epi-layers by MOCVD and Their Application to Microwave MESFETs. J. Cryst. Growth, vol. 55, no. 1, 1981, pp. 255-262.

64. Bonnet, M., et al.: Comparison of FET Performance vs. Material Growth Techniques. J. Cryst. Growth, vol. 55, no. 1, 1981, pp. 235-245.

65. Change, L. L., et al.: Semiconductor Superlattices by MBE and Their Characterization. Prog. Cryst. Growth Characterization, vol. 2, no. 1, 1979, pp. 3-13.

66. Mimura, T., et al.: A New Field-Effect Transistor with Selectively-Doped GaAs/n-GaAlAs Heterojunction. Jpn. J. Appl. Phys., vol. 19, 1980, pp. L225-L227.

67. Luschev, P. E.: Crystal Growth by MBE. Solid-State Technol., vol. 20, Dec. 1977, pp. 43-52.

68. Arthur, J. R.: Interaction of Ga and As_2 Molecular Beams with GaAs Surfaces. J. Appl. Phys., vol. 39, 1968, p. 4032.

69. Cho, A. Y.: Film Deposition by Molecular Beam Techniques. J. Vac. Sci. Technol., vol. 8, 1971, p. 531.

70. Donnelly, J. P.: Ion-Implantation in GaAs. Inst. Phys. Conf. Series, no. 33b, 1977, pp. 166-190.

71. Gibbons, J. F., et al.: Projected Range Statistics. Douden, Hutchinson, Ross, Stroudsburg, Pa., 1975.

72. Liu, S. G., et al.: Ion-Implantation of S and Si in GaAs. RCA Rev., vol. 41, no. 2, June 1980, pp. 227-260.

73. Liu, S. G. et al.: Annealing of Ion-Implanted GaAs with a Pulsed Ruby Laser. Symp. Proc. on Laser and Elec. Beam Processing of Materials. Academic Press, 1980, p. 341.

74. Pianetta, P. A., Stolte, C. A., and Hauser, J. L.: Pulsed E-Beam Ruby Laser Annealing of Ion-Implanted GaAs. Symp. Proc. on Laser and Electron Beam Processing of Materials. Academic Press, 1980.
75. Arai, M., Nishiyama, K., and Watanabe, N.: Radiation Annealing of GaAs Implanted with Si. Jpn. J. Appl. Phys., vol. 20, 1981, p. L124.
76. Long, S. I., et al.: High-Speed GaAs Integrated Circuits. Proc. IEEE, vol. 70, no. 1, 1982, pp. 35-45.
77. Taylor, G. C., et al.: GaAs Power FETs for K-Band Operation. RCA Rev., vol. 42, no. 4, Dec. 1981, pp. 508-521.
78. Gewartowski, J. W.: Progress with CW IMPATT Circuits at Microwave Frequencies. IEEE Trans. Microwave Theory Tech., vol. MTT-27, no. 5, 1979, pp. 434-441.
79. Bosch, B. G., and Engelman, R. W. H.: Gunn-Effect Electronics. John Wiley & Sons, Inc., New York, 1975.
80. Bolman, P. J., et al.: Transferred-Electron Devices. Academic Press, New York, 1972.
81. Haddad, G. I., Editor: Avalanche Transit Time Devices. Artech House, Dedham, Mass., 1973.
82. Elta, M. E., and Haddad, G. I.: High-Frequency Limitations of IMPATT, MITATT, and TUNNETT Mode Devices. IEEE Trans. Microwave Theory Tech., vol. MTT-27, no. 5, 1979, pp. 442-449.
83. Kuno, H. J.: Solid-State Millimeter-Wave Power Sources and Combiners. Microwave J., vol. 24, no. 6, 1981, pp. 21-34.
84. Hughes Aircraft Co.: Solid-State Millimeter-Wave Products. 1982.
85. Honjo, K., and Takayama, Y.: A 25-W, 5 GHz GaAs FET Amplifier for a Microwave Landing System. IEEE Trans. Microwave Theory Tech., vol. MTT-29, no. 6, 1981, pp. 579-582.
86. Dornan, B., et al.: A 4-GHz GaAs FET Power Amplifier; An Advanced Transmitter for Satellite Down-Link Communications Systems. RCA Rev., vol. 41, no. 3, 1980, pp. 472-503.
87. Snapp, C. P.: Microwave Bipolar Transistor Technology - Present and Prospects. Proc. 9th European Microwave Conf., 1979, pp. 3-12.
88. Sechi, F. N., et al.: High-Efficiency GaAs MESFET Linear Amplifiers Operating at 4 GHz. Dig. Tech. Papers, 1977 Int. Solid State Circuits Conf., Phila., Pa., pp. 164-165.

89. Liechti, C. A.: Microwave Field-Effect Transistors, - 1976. IEEE Microwave Theory Tech., vol. MTT-24, 1975, pp. 279-300.

90. Yamasaki, H., and Keithley, G. W.: High-Performance, Low-Noise FET Operating from X-Band Through Ka-Band. IEDM Dig., 1980, pp. 106-108.

91. Pucel, R., et al.: Signal and Noise Properties of GaAs Microwave FETs. Adv. in Electronics and Electron Physics, vol. 38, 1975, pp. 195-265.

92. Ruch, J.: Electron Dynamics in Short-Channel FETs. IEEE Trans. Electron Devices, vol. ED-19, 1972, pp. 652-654.

93. Van der Ziel, A.: Thermal Noise in FETs. Proc. IRE 50, 1962, pp. 1808-1812.

94. Baechtold, W.: Noise Behavior of Schottky Barrier Gate Field Effect Transistors at Microwave Frequencies. IEEE Trans. Electron Devices, vol. ED-18, 1971, pp. 97-104.

95. Statz, H., Haus, H., and Pucel, R.: Noise Characteristics of Gallium Arsenide Field-Effect Transistors. IEEE Trans. Electron Devices, vol. ED-21, 1974, pp. 549-562.

96. Fukui, H.: Optimal Noise Figure of Microwave GaAs MESFETs. IEEE Trans. Electron Devices, vol. ED-26, 1979, pp. 1032-1037.

97. Haus, H. A., and Adler, R. B.: Circuit Theory of Linear Noisy Networks. Technology Press, Cambridge, Mass., 1959.

98. Keithley, G., et al.: E-Beam Technology for K-Band GaAs FETs. Hughes Aircraft Co, R and D Rept. DELET-TR-77-2696-F, Aug. 1981.

99. Weinreb, S.: Low-Noise Cooled GaAs FET Amplifiers. IEEE Trans. Microwave Theory Tech., vol. MTT-28, no. 10, 1980, pp. 1041-1053.

100. Maloney, T., and Frey, J.: Frequency Limits of GaAs and InP Field-Effect Transistors. IEEE Trans. Electron Devices, vol. ED-22, 1975, pp. 357-358.

101. DiLorenzo, J. V., and Wisseman, W. R.: GaAs Power MESFETs: Design, Fabrication, and Performance. IEEE Trans. Microwave Theory Tech., vol. MTT-27, 1979, pp. 367-378.

102. Wemple, S. H., et al.: Control of Gate-Drain Avalanche in GaAs MESFETs. IEEE Trans. Electron Devices, vol. ED-27, no. 6, 1980, pp. 1013-1018.

103. Curtice, W. R.: Investigation of Voltage Breakdown in GaAs MESFETs with and without Gate Recess and Including Surface Depletion Effects.

Eighth Biennial Conf. on Active Microwave Semiconductor Devices and Circuits, Cornell Univ., Ithaca, N.Y., Aug. 1981.

104. Fukuta, M., et al.: Power GaAs MESFETs with High Drain-Source Breakdown Voltage. IEEE Trans. Microwave Theory Tech., vol. MTT-24, 1976, pp. 312-317.

105. Higashisaka, A., et al.: A High-Power GaAs MESFET with Experimentally Optimized Pattern. IEEE Trans. Electron Devices, vol. ED-27, no. 6, 1980, pp. 1025-1028.

106. Narayan, S. Y., et al.: X-Band Power GaAs FETs. Final Report, AFAL-TR-78-172, Nov. 1978.

107. Camisa, R. L., et al.: Lumped-Element GaAs FET Power Amplifiers. RCA Rev., vol. 42, no. 4, 1981, pp. 557-575.

108. Wemple, S. H., and Huang, H. C.: Thermal Design of Power GaAs FETs; in DiLorenzo, J. V., Editor: GaAs FET, Principles and Technology. Artech House, Dedham, Mass., 1981.

110. Sechi, F. N.: High-Efficiency Microwave FET Power Amplifiers. Microwave J., vol. 24, no. 11, 1981, pp. 59-66.

111. Cusack, J. M., et al.: Automatic Load Contour Mapping for Microwave Power Transistors. IEEE Trans. Microwave Theory Tech., vol. MTT-22, 1974, pp. 1146-1152.

112. Sechi, F. N.: Design Procedure for High-Efficiency Linear Microwave Power Amplifiers. IEEE Trans. Microwave Theory Tech., vol. MTT-28, no. 11, 1980, pp. 1157-1163.

113. Bellier, S. P., et al.: Reliability of Microwave Gallium Arsenide Field-Effect Transistors. 13th Annual Proc. Reliability Physics, 1975, pp. 193-199.

114. Abbott, D. A., et al.: Some Aspects of GaAs MESFET Reliability. IEEE Trans. Microwave Theory Tech., vol. MTT-24, 1976, pp. 317-321.

115. Irir, T., et al.: Reliability Study of GaAs MESFETs. IEEE Trans. Microwave Theory Tech., vol. MTT-24, 1976, pp. 321-328.

116. Lundgen, R. E., et al. Reliability Study of Microwave GaAs Field-Effect Transistors. 16th Annual Proc. Reliability Physics, 1978, pp. 255-260.

117. Irvin, J. C., et al.: Failure Mechanisms and Reliability of Low-Noise GaAs FETs. Bell Syst. Tech. J., vol. 57, 1978, pp. 2823-2846.

118. Mizuishi, K., et al.: Degradation Mechanism of GaAs MESFETs. IEEE Trans. Electron Devices, vol. ED-26, 1979, pp. 1008-1014.

119. Huang, C. L., et al.: Reliability Aspects of 0.5 μm and 1.0 μm Gate Low-Noise GaAs FETs. 17th Annual Proc. Reliability Physics, 1979, pp. 143-149.

120. Fukui, H., et al.: Reliability of Power GaAs FETs. IEEE Trans. Electron Devices, vol. ED-29, no. 3, 1982, pp. 395-401.

121. Slusark, W.: Reliability of Aluminum Gate C-Band GaAs Power FETs. GaAs Workshop at Reliability Physics Symp., San Diego, Calif., March 1982.

122. Van Tuyl, R. L., et al.: High-Speed Integrated Logic with GaAs MESFETs. IEEE J. Solid State Circuits, vol. SC-9, no. 5, 1974, pp. 269-276.

123. Liechti, C. A., et al.: GaAs Logic for Multi-Gb Data Generators. Int. Solid State Circuits Conf. Dig. Tech. Papers, Feb. 1982, Paper 35, pp. 172-174.

124. Eden, R. C., et al.: Multi-Level Logic Gate Implementation in GaAs ICs Using Schottky Diode FET Logic. 1980 Int. Solid State Circuits Conf. Dig. Tech. Papers, Feb. 1980.

125. Hindin, H. J., and Posa, J. G.: Gallium Arsenide Technology Forges Ahead Towards Very Large-Scale Integrations. Electronics, vol. 55, no. 4, Feb. 1982, pp. 111-126.

126. Faricelli, J. V., et al.: Physical Basis of Short-Channel MESFET Operation: Transient Behavior. IEEE Trans. Electron Devices, vol. ED-29, no. 3, 1982, pp. 377-388.

127. Greiling, P. T., et al.: Future Applications for Digital GaAs ICs. LAMBDA, 1st Quarter, 1981, p. 44.

128. Barna, A.: Advantages and Disadvantages of GaAs VLSI as Compared to Si VLSI. VLSI Design, Jan./Feb. 1982, pp. 40-41.

129. Eden, R. C.: Applicability of GaAs Integrated Circuits for Ultra-High-Speed VLSI. VLSI Design, Jan./Feb. 1982, pp. 41-43.

130. Eden, R. C.: GaAs Digital Integrated Circuits for Ultra-High-Speed LSI/VLSI; in D. F. Barbe, Editor: Very Large Scale Integration. Springer-Verlag, New York, 1980, pp. 128-174.

131. Welch, B. M., et al.: LSI Processing Technology for Planar GaAs Integrated Circuits. IEEE Trans. Electron Devices, vol. ED-27, no. 6, 1980, pp. 1116-1123.

132. Zuleeg, R., et al.: Femtojoule High-Speed Planar GaAs E-JFET Logic. IEEE Trans. Electron Devices, vol. ED-25, 1978, pp. 628-639.

133. Morkoc, H., et al.: A Study of High-Speed Normally-Off and Normally-On $Al_{0.5}Ga_{0.5}As$ Heterojunction Gate GaAs FETs (HJFET). IEEE Trans. Electron Devices, vol. ED-25, 1978, pp. 619-627.

134. Fukuta, M., et al.: Low-Power GaAs Digital Integrated Circuits with Normally-Off MESFET. IEEE Trans. Electron Devices, vol. ED-25, 1978, p. 1340.

135. Bert, G., et al.: Femtojoule Logic Circuit Using Normally-Off GaAs MESFETs. Electron. Lett., vol. 13, 1977, pp. 644-645.

136. Mizutani, T., et al.: Gigabit Logic Operation with Enhancement-Mode GaAs MESFET ICs. IEEE Trans. Electron Devices, vol. ED-29, no. 2, 1982, pp. 199-204.

137. Greiling, P. T., et al.: Electron-Beam Fabricated High-Speed Digital GaAs ICs. Proc. IEEE, vol. 70, no. 1, 1982, pp. 52-58.

138. Spicer, W. E., et al.: Fundamental Studies of III-V Surfaces and the (III-V)-Oxide Interface. Thin Solid Films, vol. 56, 1979, pp. 1-18.

139. Nuzillat, G., et al.: Quasi-Normally-Off MESFET Logic for High-Performance GaAs ICs. IEEE Trans. Electron Devices, vol. ED-27, no. 6, 1980, pp. 1102-1111.

140. Upadhyayula, L. C.: Personal Communication.

141. Yokayama, N., et al.: Low-Power High-Speed Integrated Logic with GaAs MOSFETs. Dig. Tech. Papers, 1979 Int. Conf. on Solid-State Devices, Tokyo, 1979.

142. Greiling, P. T., et al.: Electron Beam Fabricated GaAs FETs Inverter. IEEE Trans. Electron Devices, vol. ED-25, 1978, p. 1340.

143. Eden, R. C.: Introduction to Special Issue on Very Fast Solid-State Technology, Proc. IEEE, vol. 70, no. 1, 1982, pp. 3-4.

144. Pucel, R. A.: Design Considerations for Monolithic Microwave Circuits. IEEE Trans. Microwave Theory Tech., vol. MTT-29, no. 6, 1981, pp. 513-534.

145. Belohoubek, E. F.: Trends in Microwave Hybrid and Monolithic Circuits for EW Applications. Presented at 1980 DOD/AOC Symp., Anaheim, Calif., 1980.

146. Strid, E., and Reed, K.: A Microstrip Probe for Microwave Measurements on GaAs FET and IC Wafers. GaAs IC Symp. Res. Abstr., 1980, Paper 31.
147. Sechi, F. N.: Private Communication.
148. Kumar, M., Dual-Gate FET Phase Shifter. RCA Rev., vol. 42, no. 4, 1981, pp. 596-616.
149. Kroemer, H.: Heterostructure Bipolar Transistors and Integrated Circuits. Proc. IEEE, vol. 70, no. 1, 1982, pp. 13-25.
150. Narayan, S. Y., et al.: Growth and Characterization of GaInAsP and GaInAs for Microwave Device Applications. RCA Rev., vol. 42, no. 4, 1982, pp. 491-507.
151. Cappy, A., et al.: Comparative Potential Performance of Si, GaAs, GaInAs, InAs, Submicrometer-Gate FETs. IEEE Trans. Electron Devices, vol. ED-27, 1980, pp. 2158-2160.
152. Cheng, K. Y., et al.: Measurements of the Γ-L Separation in GaInAs by Ultraviolet Photoemission. Appl. Phys. Lett., vol. 40, no. 5, 1982, pp. 423-425.
153. Kajiyama, K., et al.: Schottky-Barrier Height of n-$Ga_xIn_{1-x}As$ Diodes. Appl. Phys. Lett., vol. 23, 1973, p. 458.
154. Leheny, R. F., et al.: An $In_{0.53}Ga_{0.47}As$ Junction Field-Effect Transistor. IEEE Electron Device Lett., vol. EDL-1, 1980, p. 110.
155. Bernard, J., et al.: Double Heterostructure $Ga_{0.47}In_{0.53}As$ MESFETs with Submicron Gates. IEEE Electron Device Lett., vol. EDL-1, 1980, p. 174.
156. Lile, D., et al., Int. Conf. on GaAs and Related Compounds, Vienna, Austria, Sept. 1980.
157. Shinoda, Y., et al.: InGaAsP n-Channel Inversion Mode Metal-Insulator Semiconductor Field-Effect Transistor with Low Interface State Density. J. Appl. Phys., vol. 52, no. 10, 1981, pp. 2301-2307.
158. Golio, J. M., et al.: Compound Semiconductor for Low-Noise Microwave MESFET Applications. IEEE Trans. Electron Devices, vol. ED-27, no. 7, 1980, pp. 1256-1261.
159. Liao, A. S. H., et al.: An InGaAs/Si_3N_4 n-Channel Inversion Mode MISFET. Electron Device Lett., vol. EDL-2, no. 11, 1981, pp. 288-289.
160. Wieder, H. H., et al.: Inversion Mode Insulated Gate GaInAs Field-Effect Transistors. Electron Device Lett., vol. EDL-2, no. 3, 1981, pp. 73-74.

161. Gardner, P. D.: Private Communication.

162. Tung, P. N., et al.: High-Speed Two-Dimensional Electron Gas FET Logic. Electron. Lett., vol. 18, no. 3, 1982, pp. 109-110.

163. Dingle, R., et al.: Electron Mobilities in Modulation-Doped Semiconductor Heterojunction Superlattices. Appl. Phys. Lett., vol. 33, 1978, pp. 665-667.

164. Drummond, T. J., et al.: Dependence of Electron Mobility on the Separation of Electrons and Donors in $Al_xGa_{1-x}As$/GaAs Heterostructures. J. Appl. Phys., vol. 52, 1981, pp. 1380-1386.

165. Tsui, D. C., et al.: Observation of Magnetophonon Resonances in a Two-Dimensional Electronic System. Phys. Rev. Lett., vol. 44, 1980, p. 341.

166. Stormer, H. L., et al.: Two-Dimensional Electron Gas at Differentially-Doped GaAs/GaAlAs Heterojunction Interfaces. J. Vac. Soc. Technol., vol. 16, 1979, p. 1517.

167. IEEE Trans. Electron Devices, vol. ED-28, no. 8, 1981. (Special Issue.)

168. Shank, C. V., et al.: Picosecond Nonequilibrium Carrier Transport in GaAs. Appl. Phys. Lett., vol. 38, no. 2, 1981, pp. 104-105.

169. Shur, M. S., et al.: Ballistic Transport in Semiconductors at Low Temperatures for Low-Power, High-Speed Logic. IEEE Trans. Electron Devices, vol. ED-26, 1979, pp. 1677-1683.

170. Shur, M. S., and Eastman, L. F.: Ballistic and Near Ballistic Transport in GaAs. IEEE Electron. Device Lett., vol. EDL-1, no. 8, 1980, pp. 147-148.

171. Hess, K.: Ballistic Electron Transport in Semiconductors. IEEE Trans. Electron Devices, vol. ED-28, no. 8, 1981, pp. 937-940.

172. Malik, R. J., et al.: Abstracts - 1981 Cornell Conf. Microwave Semiconductor Devices, Aug. 1981, Ithaca, N.Y.

173. Curtice, W. R., et al.: A Temperature Model for the GaAs MESFET. IEEE Trans. Electron Devices, vol. ED-28, no. 8, 1981, pp. 954-961.

174. Curtice, W. R.: Private Communication.

175. Bozler, C. O., et al.: Fabrication and Numerical Simulation of the Permeable Base Transistor. IEEE Trans. Electron Devices, vol. ED-27, no. 6, 1980, pp. 1128-1141.

176. Bozler, C. O., and Alley, G. D.: The Permeable Base Transistor and Its Application to Logic Circuits. Proc. IEEE, vol. 70, no. 1, 1982, pp. 46-52.

177. Malik, R. J., et al.: Planar-Doped Barriers in GaAs by MBE. Electron. Lett., vol. 16, no. 22, 1980, pp. 836-837. Published by the Institution of Electrical Engineers (British).

IV

Task 3: Competing Technologies

A. RF POWER GENERATION

Among the various components for a space communications system, the one responsible for the generation of the required rf transmit power in space is probably the most important and is expected to see substantial improvements in the state of the art in the next two decades.

Traveling-wave tubes (TWTs) in conjunction with fixed reflector antennas have been used nearly exclusively in all satellites up to the present. However, for the future there is a definite trend for solid-state devices to take over. Aside from the possible use of high-power Si bipolar transistors for L-band applications or Si IMPATT amplifiers at millimeter-wave frequencies, the solid-state device likely to replace most other choices is the GaAs FET. Near term we have the launch of SATCOM satellites starting at the end of 1982 that will carry a full complement of GaAs solid-state power amplifiers operating at C-band; long term we expect to see large numbers of moderate-power FET amplifiers integrated individually with antenna elements to form large phased-array structures with the promise of pushing both ERP and frequency far beyond our present limits.

1. Traveling-Wave Tubes vs Solid-State Amplifiers

TWTs have been dominating satellite communications systems from the early beginnings of Project Relay to today's sophisticated high-channel capacity satellites. Whereas low-noise TWTs have long disappeared, power tubes have shown impressive improvements over the years, from a 10-W power level at 4.2 GHz with 35-dB gain and 21% efficiency for the Relay satellite in 1962 to tubes that deliver 30 W at 12 GHz with 55 dB and 42% efficiency today. Modern helix tubes [1] with sophisticated support structures are expected to be able to handle up to 500 W cw at 30 GHz if sufficient funds are channeled into tube development over the next two decades. Even higher power levels could, in principle, be achieved with coupled-cavity-type slow-wave structures.

While the basic performance capability of TWTs has greatly improved over the years and further substantial growth can be expected, the long-term reliability of tubes is not changing much. MTBF values of nine years with 90% confidence levels were quoted as early as 1962 for Relay satellite tubes [2], although these values appear in hindsight somewhat optimistic. Today'c C-band

TWTs for the SATCOM series have typical MTBF values of six years at the 90% confidence level. Actually, more realistic calculations based on wear-out models rather than the more optimistic constant-failure-rate models predict even shorter useful life spans [3]. Although quite a bit of progress has been made over the years in cathode reliability, tube processing, and improved materials, the actual life expectancy of TWTs does not appear to get better since newer tubes operate at higher power densities and require much better controlled electron beams to permit the attainment of the quoted high efficiencies with multistage collectors. The TWT reliability problem is compounded by the poor long-life performance of the necessary high-voltage power supplies. Practically all users of power tubes in satellites cite difficulties with high-voltage supplies and poor reliability as the most serious problems facing the transmitter area.

In view of these concerns and the maturity of the TWT field, which makes major breakthroughs in reliability unlikely, it would be highly desirable to replace TWTs more and more with solid-state power amplifiers - especially if this can be done in a way that adds graceful degradation to the system by use of a large number of lower-power amplifiers.

The first step in this direction has been taken by RCA in equipping the next communications satellite, SATCOM IV, with all-solid-state power amplifiers (SSPAs), instead of TWTs.* The SSPA [4] uses 10 GaAs FETs as active devices and produces a power output of 8.5 W with 30% overall efficiency and 55-dB gain in the 3.7- to 4.2-GHz range. The choice of a solid-state power amplifier over the TWT was based on careful tradeoff studies that included not only the key parameters of size, weight, reliability, and efficiency, but also other factors such as supply-voltage requirements, linearity and, last but not least, ultimate cost. The striking advantage the SSPA has over the TWT in reliability is shown in figure IV-1, which compares the probability of 24-channel availability as a function of mission years for the two approaches. Since GaAs devices are still in a state of rapid change compared to TWTs - the first MESFETs with usable gain in the microwave frequency range appeared in 1970 - it can be safely concluded that, with the progress expected from GaAs FETs in the next decade, future satellites will increasingly use solid-state power amplifiers as fast

*Bell's Telstar is expected also to use GaAs FET power amplifiers in the transmitter.

as power devices become available at higher frequencies. Efforts are already underway to develop space-qualified GaAs FET power amplifiers for the 8-GHz communications band.

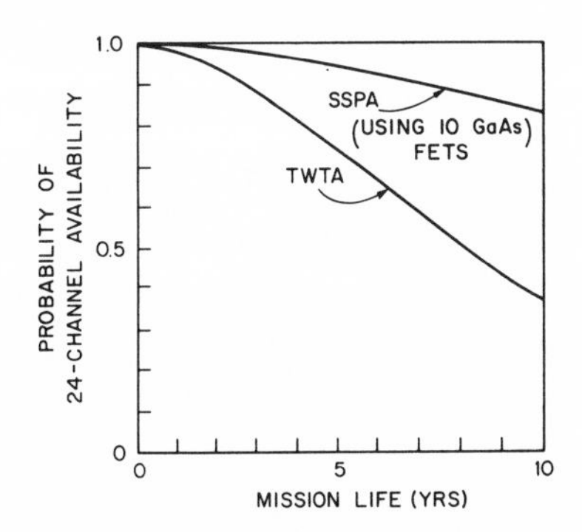

Figure IV-1. Power amplifier reliability comparison (satellite transponder 3.7-4.2 GHz).

2. Active Antenna Array

Whereas an excellent case can be made for the direct replacement of moderate-power TWTs by equivalent SSPAs based on the present state of the art of GaAs FET devices, the emphasis is expected to shift even more toward solid state in the future, when lower-cost solid-state technology will permit the use of large numbers of amplifiers directly integrated with the elements of a phased-array structure. Instead of combining many devices in one amplifier envelope to replace a given high-power TWT, it may be more cost effective to perform the power combining in space. Especially as antennas for future satellites will become more complex with a multitude of switchable spot beams, adjustable beam shapes, and low-sidelobe requirements, the phased-array concept offers significant advantages. Since rf distribution networks for large phased arrays become quite lossy, a strong case can be made for integrating the power amplifiers directly with each antenna element or groups of antenna elements. This approach is being studied at present, both under military and private company sponsorship.

RCA has a multiyear program underway to explore the possibility of using high-efficiency FET amplifiers in a phased array for future Direct Broadcast Satellite (DBS) applications. Figure IV-2 shows a conceptual drawing of such an antenna. Groups of four antenna elements are fed directly by very small, lumped-element GaAs FET amplifiers mounted on the back of the support structure of the antenna. RF and dc distribution networks are placed in one or two additional layers below the amplifiers. This arrangement provides great flexibility for amplitude and phase tapering of the antenna aperture and permits a wide latitude in beam shaping. The gain of the amplifiers and the layout of the rf feed network can be arranged so that a high degree of reliability can be achieved: a few, random failures of amplifiers have very little effect on the overall antenna performance.

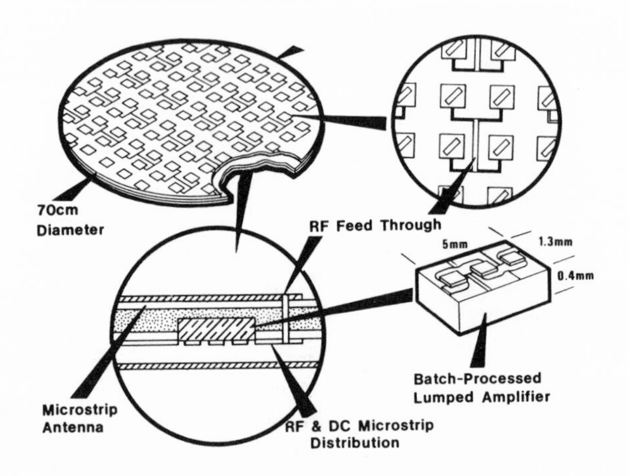

Figure IV-2. Active antenna array.

By splitting the total required antenna power output into many low-power individual amplifiers, a series of additional advantages is gained. First, the efficiency of a lower-power device is higher than that of a larger, higher-power device. This difference is expected to diminish as GaAs technology matures but it remains a factor nevertheless. Second, when more lower-power amplifiers are being used, fewer antenna elements have to be combined per amplifier, and this leads to reduced rf distribution loss. Third, by operating a high-power device at reduced power output, the operating temperature of the device drops and the effective life span is increased. Recent experiments performed at RCA showed, for example, that a 0.5-W Microwave Semiconductor Corp. (MSC) device operating at the 150-mW level at 12 GHz produced a power-added efficiency of 38%, compared with 28% at full power. The high-efficiency operation results in a drop in operating temperature by 20°C and consequently increases the expected lifetime by more than an order of magnitude.

Typical trends for the overall efficiency of multistage GaAs FET amplifiers, for different power levels at 12 GHz, are given in figure IV-3. These curves are intended to show the expected performance characteristics of complete amplifiers suitable for space use and are thus rather conservative with respect to already achieved state-of-the-art "best performance" values.

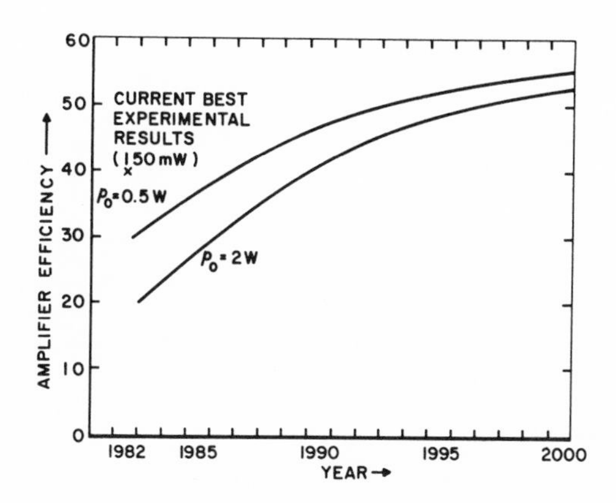

Figure IV-3. Projection of overall amplifier efficiency using GaAs FETs at 12 GHz.

Based on these efficiency figures together with the previously mentioned advantages of better flexibility in radiation pattern shaping and lower voltage-supply requirements, the phased-array system in which large numbers of individual amplifiers are integrated directly with antenna elements appears to be superior to the conventional reflector antenna with a single TWT power source, even when the most sophisticated depressed collector schemes are used. The main feature detracting from this favorable picture is the higher cost associated with the larger number of amplifiers. Here, however, extrapolations of GaAs device costs based on similar trends in Si devices, together with the promise of an eventual development of a batch process solid-state circuit technology, indicate that the solid-state approach very well may also become more cost effective than the tube approach.

The use of individual low-power amplifiers for small groups of antenna elements may also permit a more advantageous dc supply arrangement for the amplifiers. As shown in figure IV-4, the power amplifiers could be connected directly to appropriate solar panels on the back of the antenna array. Each amplifier thus operates with full autonomy, providing better reliability and independence from a large central dc power distribution network.

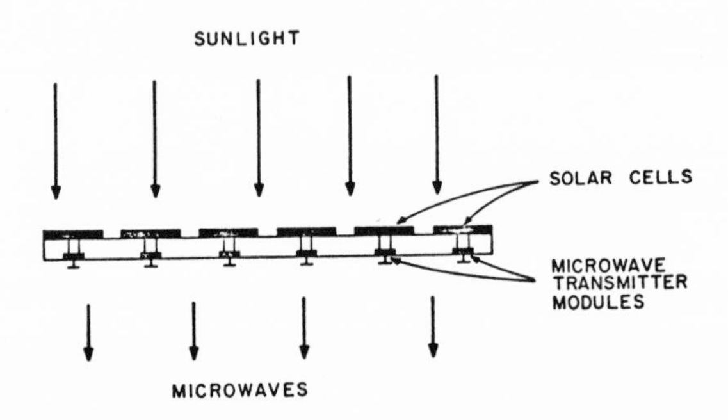

Figure IV-4. Direct integration of amplifiers with solar panels.

Since a satellite in synchronous orbit receives the sunlight from different directions in the course of its path, a separate focusing arrangement for the sun rays is necessary. Figure IV-5 shows the concept of a possible rotating-mirror arrangement that achieves this effect. A stationary mirror or focusing structure is attached to the back of the antenna array under an angle of 45°. A second rotating mirror ensures that the sunlight is focused on the back of the array at all times. The light intensity available at the back of the array without any additional concentrating schemes is enough to provide the necessary dc power for 100-mW amplifiers, each of which feeds groups of four antenna elements spaced at 0.8 λ at 12 GHz. This would be sufficient to provide

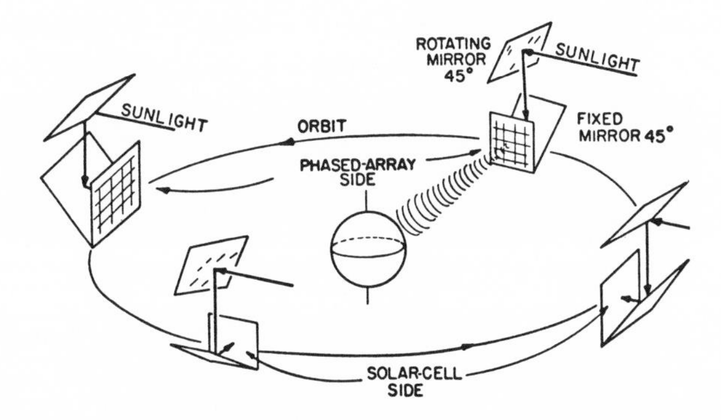

Figure IV-5. Proposed arrangement for focusing sunlight onto back of antenna array.

a total radiated power of 200 W, as required for a direct-broadcast-satellite antenna having a diameter of approximately 2 m.

3. Comparison of Solid-State Sources

The advantages of solid-state amplifiers appear to be fairly clear-cut as long as GaAs devices with sufficient output power and low cost are available. If, as is to be expected, frequencies above 20/30 GHz will be required by the year 2000 to handle the increased channel and bandwidth requirements, alternate solutions may be called for. Aside from TWTs, IMPATT amplifiers and possibly InP Gunn effect amplifiers may be considered as alternate approaches.

The first IMPATT devices for microwave amplifiers appeared just a few years before the onset of the GaAs FET revolution, and early IMPATT amplifiers were considered a viable solid-state replacement for TWTs in communications systems. 1-W IMPATT amplifiers for digital communications systems at 11 GHz, which became available in the early 1970s, caused very little deterioration in the bit error rate when operated under 40-Mbit QPSK modulation conditions [5]. Bell Laboratories developed a 3-W IMPATT amplifier for their 11-GHz FM radio relay system. At present, IMPATT amplifiers for this system are still manufactured [6] on a replacement basis and will continue to be used for the foreseeable future.

The main advantage of IMPATT devices is their high power-output capability compared to that of GaAs FET devices. This becomes especially evident for frequencies above 20 GHz. By combining 12 diodes in one waveguide cavity, TRW recently demonstrated a 10-W IMPATT amplifier at 41 GHz with 30-dB gain and a bandwidth of 250 MHz for a spacecraft communications transmitter [7]. The major findings of this development are that aside from the successful parallelling of many diodes, the error bit rate even for injection-locked operation (which provides much higher gain than the regular amplifier mode) is not much affected, showing a signal-to-noise degradation of less than 1 dB. With the IMPATTs' inherent capability to operate at substantially higher frequencies (e.g., close to 1 W cw has been obtained experimentally from a single device by Hughes Aircraft Co. at 94 GHz), they do indeed represent a viable power source for millimeter-wave communications systems.

However, IMPATT amplifiers also have a number of inherent drawbacks that are unlikely to change. Their gain-bandwidth product is generally rather limited, and difficulties with parametric and subharmonic oscillations make them hard to adjust for long-time reliable performance. Bias-instability problems become especially pronounced with the more efficient Read profile devices. One of their major drawbacks, for which no reasonable solution is expected, is their inherent nonlinearity, which causes them to show great differences in gain-bandwidth as a function of drive level. Whereas this behavior is tolerable in present FM systems, such as Western Electric's 11-GHz radio relay system, future applications with their strong requirements for increased bandwidth and ultralinear operation to sustain modern modulation schemes are not compatible with IMPATT characteristics.

Another alternate solid-state amplifier source for mm-wave frequencies is the Gunn-effect amplifier. This device received considerable attention in the time period 1965-1973, before GaAs amplifiers with equal or better performance characteristics became available. In principle, these amplifiers can offer stable wide-bandwidth, linear, high-gain operation at frequencies well into the mm-wave range. For example, InP Gunn-effect amplifiers have shown power outputs of 100 mW at 50 GHz with 20% bandwidth and 6-dB gain per stage [8]. However, with all the effort that went into amplifiers of this type in previous years, they remain limited in three major areas: noise figure, power output, and efficiency. Although presently capable of operating at much higher frequencies than GaAs FETs, Gunn-effect devices are not expected to play a major role in medium-power amplifiers for future satellite communications systems, mainly because of their low efficiency. At frequencies where both Gunn effect and GaAs FETs can be used, the FET has the clear advantage of better linearity and a higher efficiency by at least a factor of 5, which increases at lower frequencies to a factor of 10 to 1 or more. Thus, the choice for the transmitter power source at higher mm-wave frequencies is expected to be between TWTs and FETs, depending on how high a frequency the FET will be able to attain.

Not all of the future system developments will necessarily occur at mm-wave frequencies. There may also be a strong need for high-power, high-efficiency transmitters at L-band or below for various forms of mobile communications systems. Aside from high-power spaceborne transmitter requirements that fall into the previously discussed tradeoff between high-power TWTs and active solid-state phased-array structures, there will possibly be a strong demand for extremely lightweight, small-size, medium-power, very high efficiency transmitter amplifiers in personal communication units.

At frequencies below S-band, the choice between GaAs and Si is not as clear-cut as at higher frequencies, and probably will be determined only by the ultimate cost of the particular configuration.

For brute power generation at frequencies in the vicinity of 1 GHz, the Si bipolar is clearly the best choice at present and probably will remain so for a long time to come. Although newer high-power structures such as vertical FETs may challenge the bipolar, it will be very difficult to produce better and more-cost-effective devices than the Si bipolar, which has been in a mature state of development for a long period of time. These devices are being fabricated today in large quantities for the mobile communications market and most likely will be able to fulfill all upcoming high power needs in the lower frequency range.

For certain specialized applications, however, such as very high efficiency, moderate-power transmitters for personal lightweight communications, GaAs is probably a better choice. For example, a power-added efficiency of 72% at the 1-W output power level has been demonstrated with GaAs FETs [9] at 2.45 GHz. At lower frequencies, even higher efficiencies (on the order of 80%) could be expected from amplifiers operating in a Class-E switching mode [10]. GaAs FETs offer the advantage of a higher cut-off frequency than that of bipolar transistors. This characteristic is essential for the proper current and voltage waveform shaping to achieve high efficiency and also provide high gain, permitting the development of very compact lightweight amplifiers with a minimum number of stages. The exceptionally high efficiency of these amplifiers makes them highly suitable for portable personal transmitters, as required in a direct person-to-satellite communications system.

B. FILTER STRUCTURES

Today's high-performance filters that are used in spaceborne transponders nearly all share the same requirements for a high-quality factor (Q_o): light weight and small size. In conventional microwave filters there is a fundamental tradeoff to be observed between Q_o and size. A high Q_o can be realized without difficulty in waveguide structures but only at the expense of size and weight. Microwave integrated-circuit (MIC) filters are much smaller in size but are also much lossier than their waveguide counterparts. Dielectric-resonator and active lumped-element filters under development now hold considerable promise to overcome the present stalemate between size and quality factor.

1. Standard Passive Filters

A typical state-of-the-art filter for channelized multiplexer applications is shown in figure IV-6. The filter uses dual-mode TE_{113} cavities with Q_o values in excess of 11,000 to provide an eight-pole quasi-elliptical response. A circulator together with an additional reflector cavity section provides group-delay equalization. Excellent temperature stability is ensured by the use of Invar* for the cavity walls. More recently, graphite-fiber-reinforced plastics instead of Invar have been used to bring the weight down. However, the size of these types of filters is more or less fixed. Coaxial and MIC filters can be made much smaller and lighter, but their Q values are one to two orders of magnitude lower, preventing a reasonably low loss performance with high skirt selectivity. There thus exists a great incentive for the development of new filter types that combine small size with high Q_o.

2. Dielectric Resonators

Resonators with moderately high Qs can be realized in small size if a high-dielectric-contant material is used to fill the cavity. Better yet,

*Nickel-ferrite controlled-expansion alloy (Precision Metals Services, Colmar, Pa.)

Figure IV-6. Eight-pole 12-GHz waveguide filter.

dielectric resonators, which confine the electromagnetic field within the dielectric without the use of conducting cavity walls, have shown excellent, very high Q performance and have found wide application in small-size, stabilized solid-state oscillators. Table IV-1 shows the properties of currently available, commercial high-ε materials. A very important feature of the new dielectric materials is their adjustability of the temperature coefficient. By proper choice of the material's composition, temperature coefficients with arbitrary positive or negative values can be obtained; these can be used to compensate for the temperature coefficients of the housing and coupling structure to provide very high overall frequency stability against temperature variations.

The last column in table IV-1 indicates approximate volume requirements for a single resonator at 10 GHz if fabricated in different media. The size for the dielectric resonators includes the necessary metal shielding to confine stray fields. The Q_o of the comparable waveguide resonator refers to a standard

TABLE IV-1. DIELECTRIC RESONATOR MATERIALS

Type	Manufacturer	ε_r	tg δ	Q_0	Temp. Coeff.	Size for 10-GHz resonator
Ba_2TiO_G	Trans-Tech Inc.	37	$.5 \times 10^{-3}$	2000	-25 PPM/°C	0.8 cm^3
R-04C Series	JFD Components (Murata)	37	1.6×10^{-4}	6300	Adjustable to 0 PPM/°C	0.8 cm^3
R-09C Series	JFD Components (Murata)	90	10^{-3}	1000	Adjustable to 0 PPM/°C	0.44 cm^3
Comparable Waveguide Rectangular Resonator		1	-	7000	Adjustable to 0 PPM/°C	6 cm^3

TE_{011} cavity. Higher-order-mode cavities can provide higher Q values, but also require a still larger volume. In general, one may conclude that the volume of dielectric resonators is approximately one-tenth that of comparable waveguide resonators.

Based on the rapidly spreading use of dielectric resonators [11,12], further improvements in the material properties can be expected that are likely to make dielectric resonators the preferred building blocks for the complex channelized filters required for the input and output multiplexers of space-borne transponders.

3. Active Microwave Filters

Looking at future, more sophisticated satellite transponders, the need for still smaller sizes and in certain cases rapid tunability or adjustability of the filter structures becomes apparent. A new active tunable microwave filter under development for the Navy [13] may very well become an important component for future spaceborne filters. The present effort is aimed at bandpass filter structures that can be rapidly tuned over a frequency of several gigahertz at X-band while offering low loss and good skirt selectivity.

Fast tuning on the order of tens of nanoseconds can readily be achieved with varactors. At microwave frequencies the Q_o of even the best varactors ramains well below 100. To overcome the losses in the varactor and the lumped elements forming a tunable filter section, a negative resistance is added in the form of a GaAs FET in a suitable feedback configuration as shown in figure IV-7. The resulting resonator element is very small, measuring approximately 10x5x5 mm^3 at X-band; its resonance frequency can be adjusted by the varactor voltage, while its Q_o can be set by selecting an appropriate gate voltage for the FET. An experimental version of a two-stage X-band filter is shown in figure IV-8. The unit has a 3-dB bandwidth of approximately 35 MHz and can be tuned over a 2-GHz frequency range.

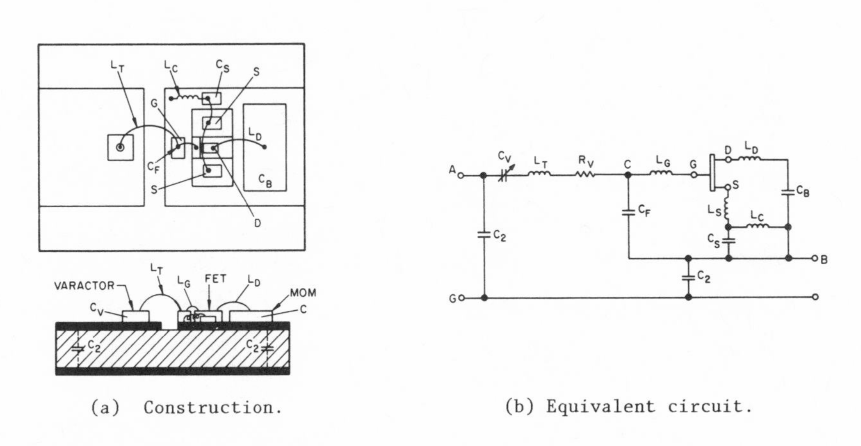

(a) Construction. (b) Equivalent circuit.

Figure IV-7. Active tunable resonator element.

The major advantages of this filter type are very small size and weight, suitability for monolithic integration, adjustable high Q_o, rapid tunability, and the possibility of being programmed by computer-controlled D/A's. The addition of an active element however, also introduces some drawbacks; these are limited dynamic range, moderately high noise figure, and the requirement of dc power for the active device.

For applications where tunability is not required, the active-filter concept can provide resonators of extremely small size compared to conventional

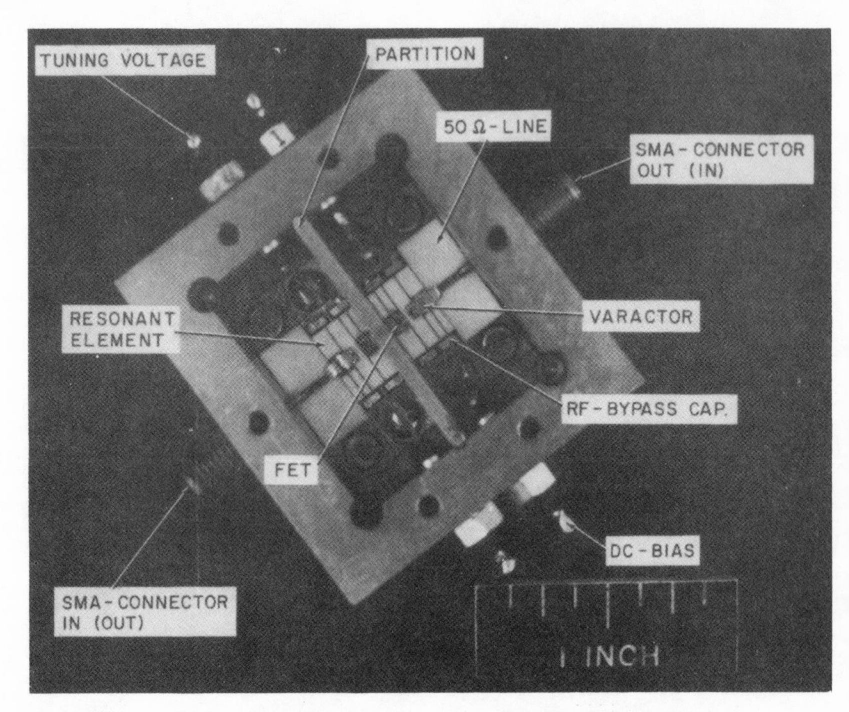

Figure IV-8. Experimental two-stage tunable X-band filter.

technologies. For example, figure IV-9 shows a single-resonator passband filter at 800 MHz that has a 3-dB bandwidth of 400 kHz, and measures only 3x8x5 mm^3. Since this type of filter lends itself very well to monolithic integration, it opens the way for the eventual fabrication of complex channelized filters having very high performance-to-volume ratios.

C. MICROWAVE CIRCUIT FABRICATION

Some consideration shall be given here to the question of what type of circuit technology is most likely to be used in future space communications systems. At present, two major choices exist: One is the standard hybrid

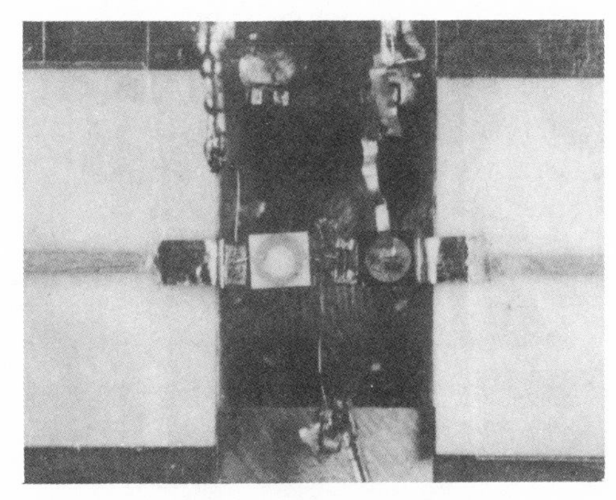

Figure IV-9. Single-element, fixed-frequency active filter (f_o = 800 MHz; 3-dB bandwidth, 400 kHz).

technology used for many years in various forms; the other, so far only in the development stage, is the monolithic technology (which has received substantial government and commercial support in the last few years). There is, in addition, a third technology; this combines several of the desirable features of the other two technologies and may be well suited for certain space applications where high efficiency and minimum size and weight are key requirements. Before attempting to predict which one of the technologies is best suited for a particular application, and in what time frame, the basic features of the three approaches will be highlighted first.

1. Conventional Hybrid Circuits

In this technology, which has been the mainstay of microwave circuit fabrication in the past, circuits are formed on various substrates by either thick- or thin-film processes; the active devices, which generally are packaged separately, are inserted at appropriate locations and interconnected with the circuits by soldering or wire bonding. The key features of this technology are high versatility, ease of trimming (essential for achieving the best possible performance from state-of-the-art devices), and fast-turnaround time for the fabrication of a moderate number of units. The major drawbacks are the relatively large size of the components and the high labor content in assembly and trimming.

The hybrid circuit technology of today is rather mature and can be implemented at different levels of sophistication and complexity, from simple amplifier circuits on Duroid* with packaged devices to complete transmit/receive modules with carrier-mounted devices and circuits fabricated on alumina or sapphire. A typical example of this technology is the multistage GaAs FET amplifier [4] shown in figure IV-10 which is earmarked to become the replacement of the currently used 4-GHz TWT amplifier in space applications. The hybrid fabrication technology used for this amplifier offers very high reliability and, with the help of individual circuit trimming, is capable of extracting the best state-of-the-art performance from present devices. Although this approach is not very economical in area usage, it requires less volume and weight than does a comparable TWT.

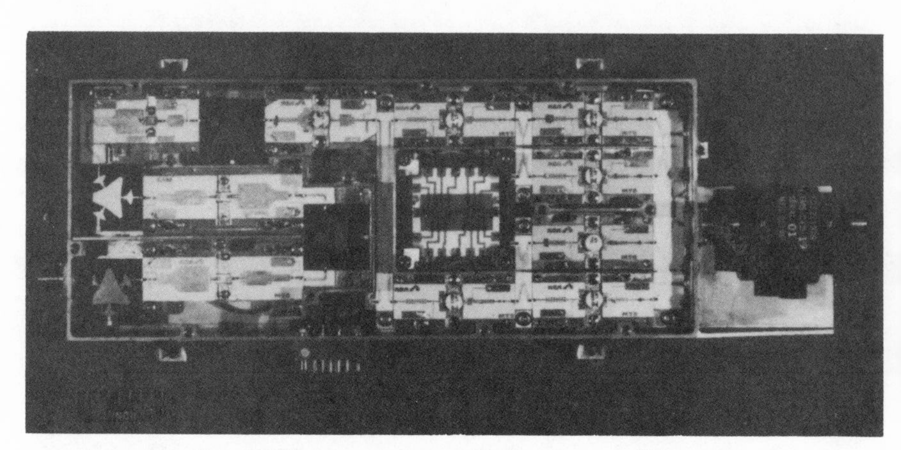

Figure IV-10. Solid-state power amplifier for satellite transponder.

*Glass-fiber coated PTFE (Rogers Corp., Chandler, Ariz.).

2. Monolithic Circuits

During the last few years there has been a widespread effort, involving a multitude of companies, to expand monolithic technology from the lower frequencies where it dominates both digital and, to a lesser extent, linear applications to the microwave frequency range. A monolithic circuit is defined as one that includes both active devices and associated circuitry on the same substrate. By eliminating the need for separately interconnecting each device with the circuit and by batch-fabricating many circuits and devices at the same time, a considerable labor savings can, in principle, be achieved. This is, however, counteracted by the larger required area of expensive GaAs real estate, more complex process steps, a lower yield, and the high degree of difficulty or even inability to trim and fine-tune such circuits. The success of the monolithic approach in the long run will be determined mostly by economic considerations. This means that only in applications where truly very large numbers of circuits are required and where some derating of the device characteristics is acceptable will monolithic circuits offer definite advantages. Thus it is highly desirable to have circuits that can be universally used and that have a small ratio of circuit to device area to take full advantage of the cost savings of the batch-fabrication process.

The above considerations lead to the conclusion that monolithic circuits will be best suited for frequencies above 20 GHz, where the circuit portion of the chip becomes reasonably small. Here, direct interfacing between circuit and device on the same chip also helps greatly to reduce interconnection parasitics, which otherwise become very difficult to control at higher frequencies. For frequencies in the vicinity of 10 GHz and lower, standard distributed circuits become rather large even on GaAs substrate material with its high dielectric constant of $\varepsilon_r = 13$. Here, lumped-element matching networks using discrete inductors and capacitors are preferred, as they occupy a much smaller area than distributed circuits. These elements require, however, more complex processing steps and also are, in general, more lossy than distributed circuits.

At the other end of the microwave spectrum, around 1 GHz, even lumped-element circuits become too large for economic use in monolithic circuits. To overcome this difficulty, some amplifiers are being developed in which certain circuit components are replaced by active devices. A good example of this approach is the replacement of passive power splitters, which are generally very space consuming, by active-device power dividers, as implemented in the monolithic balanced mixer design of Plessey [14]. Another approach is based on using active loads and active matching elements, as demonstrated by the wide-band instrumentation amplifier [15] of Hewlett-Packard. This amplifier covers the frequency range from 5 MHz to 3 GHz with 50-dB gain, while measuring only 0.25 mm x 0.25 mm in area. It is thus an excellent example of what monolithic circuits can do best: provide low-power, wide-bandwidth performance in a very small chip size.

The main drawbacks of the two latter approaches are higher dc dissipation due to the use of additional active devices and, generally, a poorer noise figure. They are also not suited for high-power applications. Thus, aside from economic considerations, monolithic amplifiers have to be matched carefully to the technical requirements to be successful.

3. Miniature Hybrid Circuits

In this technology, very small lumped-element circuits are batch processed, similar to those used in the monolithic approach, the only difference being that the active devices are later attached to the completed and pretested circuit substrates. Whereas the circuit size is thus comparable with that of monolithic circuits, active devices are processed separately on substrates tailored to device performance rather than circuit requirements. Since in most cases the substrate requirements for circuits and devices are quite different with different processing procedures, this separation of device and circuit permits a better optimization for either type. Devices are mounted separately and are being screened before mounting, thus increasing the overall yield. The separation of circuit and device substrates also leads to better heat dissipation and lower rf losses.

Typical early examples of this technology are the miniature low-noise and low-power amplifiers manufactured by Avantek. Individual amplifier stages are made of batch-processed circuit boards with device chips subsequently mounted and bonded to the circuits. A 9-stage amplifier having 44 dB of gain over the frequency range from 12 to 18 GHz covers an area of only 0.350 in. x 2.300 in., including the housing. More recently, Watkins-Johnson came out with the so-called "Super FET" [16], which combines two FETs with an interstage matching network on a single piece of GaAs. The resulting device requires standard input and output matching networks, much as does a regular FET; it has, however, the advantages of much better gain and excellent isolation and stability, and is only somewhat larger than two separate FETs. The "Super FET" combines advantages of monolithic technology very effectively with hybrid fabrication.

Another approach being pursued by RCA, especially for high-power solid-state amplifiers, consists of processing all circuit components of a multistage amplifier on a common BeO substrate with integrated septa for low-inductance ground returns at the points where the active devices are separately attached. Figure IV-11 shows the basic concept of this technology [17]. Its main advantage is excellent heat conduction through the BeO substrate, while providing a high-Q support substrate for the lumped-element matching circuits. In addition, the thermal expansion coefficient of BeO matches closely that of GaAs, thus permitting a high-quality bond between the two materials, with a minimum of mechanical stress.

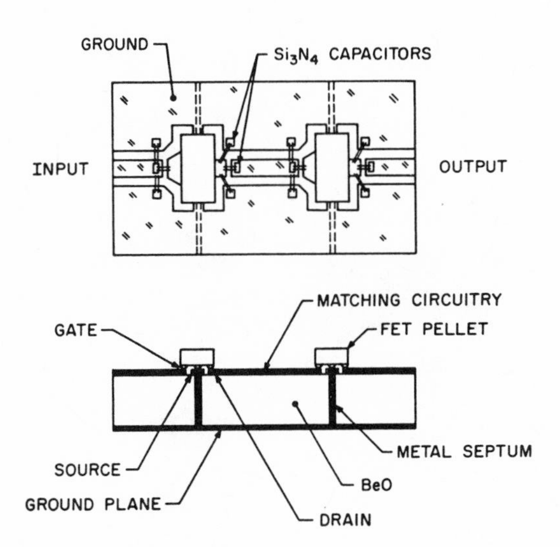

Figure IV-11. Schematic diagram of two-stage miniature beryllia amplifier.

Devices are preferably bump-mounted by a method similar to the one that is widely used at lower frequencies where large-area LSI chips are mounted by solder reflow of a multiplicity of contact bumps. This technology provides the mechanical mounting of the chip with good thermal dissipation properties simultaneously while making all the necessary electrical connections. It eliminates the need for separate wire bonds, which generally are a source of poorly reproducible parasitic elements and require costly, highly skilled assembly labor.

The very high efficiency results presented in figure IV-3 were obtained with a miniature BeO amplifier of the type described above. The active area of the 12-GHz single-stage amplifier measures only 5 mm x 3 mm. A photograph of the amplifier is shown in figure IV-12.

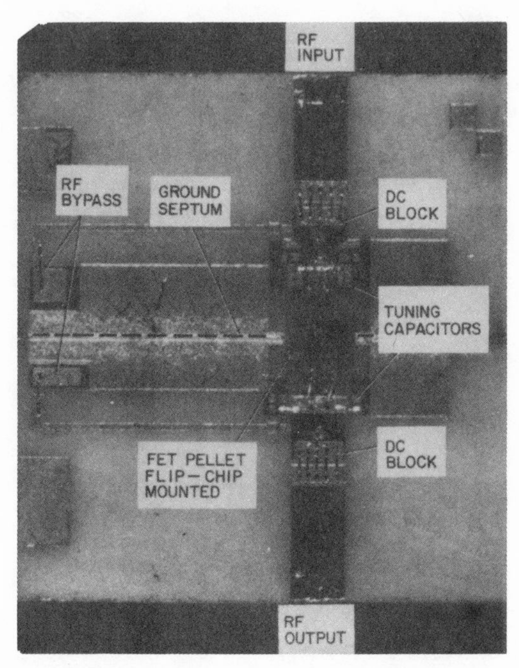

Figure IV-12. 12-GHz lumped-element high-efficiency amplifier.

Although many of the above-mentioned technologies and processes are still experimental today, they promise to greatly simplify the currently used common hybrid-circuits technology and to provide a means for moderate volume fabrication without going all the way to a monolithic technology.

4. Discussion and Outlook

No single fabrication technology can be best suited to all applications. One trend, however, appears to be reasonably clear: standard hybrid circuits, in the form in which they are widely used today, will become less and less attractive for use in future satellite communications systems. They will be replaced nearly everywhere by either monolithic or miniature hybrid circuits.

For spaceborne applications, the major decisive factors in selecting a particular technology are performance and size/weight. Cost is here secondary to performance parameters, such as efficiency, noise, distortion, and reliability. For ground applications, the emphasis is reversed. Cost becomes more critical since it is borne directly by the individual end user, as for example, ground terminal receivers for the DBS system.

In general, monolithic circuits can be expected to be best at handling very high volume, low- or medium-power applications in which size and weight are the most important factors. Monolithic circuits will be more cost effective at higher microwave and millimeter-wave frequencies where the rf circuit becomes an ever smaller part of the total chip. Area reductions by two or three orders of magnitude compared to present conventional hybrids can be achieved. Examples of promising large-volume applications are the rf portion of Earth terminal receivers, transmit/receive modules for spaceborne phased arrays, especially above 20 GHz, and certain portions of personal lightweight communication modules. Monolithic circuits are generally less suited for high-power or lower-frequency applications and in situations where the very best state-of-the-art, narrowband (less than 20%) performance has to be extracted from active devices.

The latter requirements are better handled by miniature hybrids. These circuits have size advantages similar to those of the monolithic counterparts. However, since the circuit portion is fabricated separately from the devices on relatively low cost ceramic substrates, it can be cost effectively produced even at lower frequencies, at which the circuit size becomes rather large compared with the device size. For high-power applications, BeO can be used as substrate material that offers excellent heat dissipation properties. Devices will increasingly be flip-chip mounted, a technique that combines low assembly cost with excellent heat conduction and low rf loss. Devices equipped with suitable bumps for flip-chip mounting are being fabricated by Mitsubishi, MSC, and RCA, and are expected to become more widely available as time passes. Miniature hybrid circuits in general permit some form of pretesting and trimming, provide lower rf losses, and are capable of better performance than comparable monolithic circuits. They are expected to find widespread use in spaceborne applications where the combination of very small size/weight and high performance is of the greatest importance. They are also expected to play a significant role at low frequencies in high-efficiency transmitter amplifiers, like the ones needed for the ultralightweight wrist-band personal radiophone.

Standard hybrid circuits, on the other hand, may still remain in favor for certain specialized high-performance ground-terminal uses where size/weight considerations are unimportant and a small production volume does not justify the development costs of other approaches.

D. REFERENCES

1. Fleury, G., Kuntsmann, J. C., and Maloney, E. D.: New TWT Generation Pumps Power for Today and Tomorrow. Microwave System News, vol. 12, no. 1, Jan. 1982, pp. 97-118.
2. Vaccaro, F. E., Wakefield, P. R., and Schindler, M. J.: Recent Advances in Traveling-Wave Tubes for Communications Satellites. RCA Engineer, vol. 10, no. 9, Dec./Jan. 1965, pp. 22-25.
3. Behmann, F. F.: Improved Reliability for New Satellite Systems. Proc. Annual Reliability and Maintainability Symp., Jan. 1981, pp. 408-413.
4. Dornan, B., et al.: A 4-GHz GaAs FET Power Amplifier: An Advanced Transmitter for Satellite Down-Link Communication Systems. RCA Rev., vol. 41, no. 3, Sept. 1980, pp. 472-503.
5. Paik, S. F.: IMPATT-Diode Power Amplifiers for Digital Communicating Systems. IEEE Trans. Microwave Theory Tech., vol. MTT-21, no. 11, Nov. 1973, pp. 716-720.
6. Irwin, C. G.: Building 800 + IMPATT Amplifiers. Microwave J., vol. 22, no. 1, Jan. 1979, pp. 57-61.

7. Mooney, D., and Bayuk, F. J.: 41 GHz 10-W Solid State Amplifier. 11th European Microwave Conf., Amsterdam, Holland, Sept. 1981, pp. 876-880.
8. Fank, F. B.: InP Emerges as Near-Ideal Material for Prototype mmW Devices, Microwave System News, vol. 12, no. 2, Feb. 1982, pp. 59-70.
9. Sechi, F. N.: High Efficiency Microwave FET Power Amplifiers. Microwave J., vol. 24, no. 11, Nov. 1981, pp. 59-66.
10. Sokal, N. O.: Class E - A New Class of High Efficiency Tuned Single-Ended Switching Power Amplifiers. IEEE J. Solid-State Circuits, vol. 10, no. 3, June 1975, pp. 168-176.
11. Stiglitz, M. R.: Dielectric Resonators: Past, Present and Future. Microwave J., vol. 24, no. 7, July 1981, pp. 19-36.
12. Atia, A. E., and Bonetti, R. R.: Generalized Dielectric Resonator Filters. Comsat Tech. Rev., vol. 11, no. 2, 1981, pp. 321-341.
13. Presser, A.: Varactor-Tunable, High-Q Microwave Filter. RCA Rev., vol. 42, no. 4, Dec. 1981, pp. 691-705.
14. Pengelly, R. S., et al.: A Comparison Between Actively and Passively Matched S-Band GaAs Monolithic FET Amplifiers. IEEE MTT - Symp., Los Angeles, Calif., June 1981, pp. 367-369.
15. Estreich, D.: A Wideband Monolithic GaAs IC Amplifier. IEEE-ISSCC Dig., vol. 25, Feb. 1982, pp. 194-195.
16. Crescenzi, E. J.: A Monolithic Device for High-Gain Amplifiers. Microwaves, vol. 20, no. 7, July 1981, pp. 65-69.
17. Sechi, F. N., et al.: Miniature Beryllia Circuits - a New Technology for Microwave Power Amplifiers. RCA Rev., vol. 43, no. 2, June 1982, pp. 363-374.

V

Task 4: Fundamental Limits

The solid-state devices discussed in this section fit functionally into two categories -- linear and logic devices. Linear devices include both low-noise and high-power devices for transponder applications; logic devices are mostly for baseband signal processing.

For future applications, the transponder carrier frequency is likely to be higher than it is now - 20 GHz and above - because of the crowded lower-frequenc spectrum. On the other hand, the trend of baseband signal processing may be expected to head toward complex, multifunctional processing to facilitate on-board switching functions.

The solid-state devices that will satisfy these trends are likely to be in the monolithic format, perhaps with medium-scale integration (MSI) for power application and large-scale integration (LSI) for signal processing. Microwave monolithic integrated circuits (MMICs) have the potential advantages of small size, light weight, low cost, and high reliability. MMICs will have the additional advantage of higher performance at higher frequencies than the same devices in the discrete format. At higher frequencies, the device impedance is, in general, very low, on the order of a few ohms. The ability of the MMIC to provide impedance matching very close to the active device leads not only to a wide operating bandwidth but also to a potential loss reduction in the impedance-matching circuit.

Having recognized the future trends of the solid-state devices, we concentrated our study on the fundamental limits in high-frequency MMIC devices and in high-density logic devices. The fundamental physical limits such as electrical, thermal, and physical-dimension limitations are presented. Research areas in new device concepts as well as technology development that will affect the limits are also identified.

A. ELECTRICAL LIMITATION OF DEVICE OUTPUT POWER AT HIGH FREQUENCIES

One of the major differences between a solid-state device and a vacuum tube is that the carrier velocity of a solid-state device reaches some saturated velocity under high electric field (high potential) conditions while the electron velocity in vacuum tubes continues to increase with applied potential.

To maintain the same carrier transit time at high frequencies, the dimensions of a solid-state device should be roughly proportional to the wavelength or inversely proportional to the operating frequency. This dimensional requirement applies to the transit-time devices such as the Gunn and IMPATT diodes, and also to the high-frequency, cutoff devices such as the bipolar and field-effect transistors.

Physical dimension is not the only limitation to the device's high-frequency performance. Even if we could make the physical dimensions as small as we should like, the small physical dimension (L) cannot support a high dc bias voltage (V) because the breakdown electric field ($\sim V_B/L$) of a semiconductor material is approximately constant, of the order of 10^5 V/cm. Thus, $V \propto L \propto 1/f$. The maximum applied voltage not causing high field breakdown is inversely proportional to the operating frequency.

On the other hand, the maximum current of a semiconductor device is proportional to Q/τ, where Q is the total available mobile charge and τ is the transit time. Since the total available charge is equal to qnAL (where n is the carrier concentration), which is proportional to qnA(1/f), the maximum semiconductor device current is also inversely proportional to the device cut-off frequency. The above considerations lead to the well-known conclusion that the device maximum output power is inversely proportional to the square of the device cut-off frequency.

This observation was first derived by E. O. Johnson of RCA Laboratories [1]. Although his derivation was intended only for the bipolar transistor, we will, in this section, derive a general equation that applies to all semiconductor devices.

1. Voltage-Frequency Relationship

The cut-off frequency, f_T, of a semiconductor device is defined as

$$f_T = (2\pi\tau)^{-1} \tag{1}$$

where τ is the transit time of a carrier passing through the rf-modulation region. In the case of transit-time devices such as Gunn and IMPATT diodes, τ is the time of flight between the two ohmic contacts. In the case of a transistor, τ is the time of flight through the gate or the base region.

The transit time τ can be calculated from the following equation:

$$\tau = \frac{L}{\upsilon_a} \tag{2}$$

where L is the distance of the rf-modulation region through which the carrier is traveling, and υ_a is the average velocity of the carrier in L. It is obvious that for high-frequency operation the average carrier velocity should be as high as possible to keep τ small.

The average carrier velocity is determined by the semiconductor velocity-electric field (V-E) characteristics and the spatial electric field distribution, which is governed by the location of the electrodes and the applied dc bias voltages. Figure V-1 shows the V-E characteristics of several types of semiconductors. High average velocity can be achieved in a semiconductor that has a high saturated drift velocity and high carrier mobility. When the mobility is high, the carrier can be accelerated to the saturated velocity within a relatively short distance and in a relatively short time, leading to an effectively high average velocity.

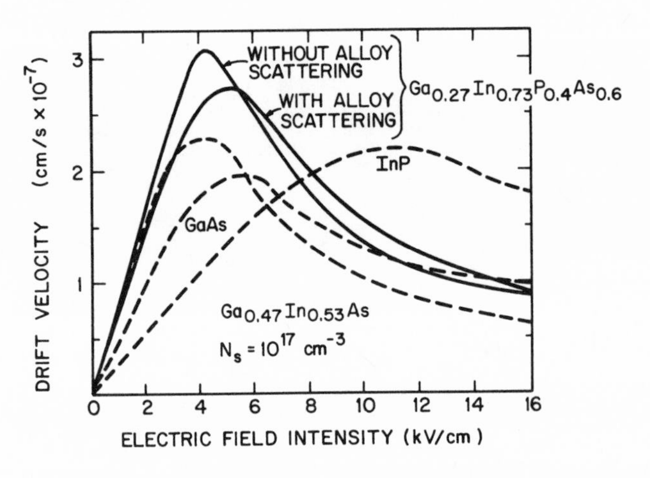

Figure V-1. Velocity-field characteristics of various semiconductors.

The transit time τ can also be reduced by decreasing the rf-modulation distance L. The lowest limit on L that will sustain a maximum voltage V_m is determined by the semiconductor breakdown field,

$$E_B = V_m/L \tag{3}$$

Figure V-2 shows the breakdown field vs carrier density for GaAs and silicon. Combining equations (1), (2), and (3), we have:

$$V_m f_T = \frac{E_B \upsilon_a}{2\pi} \tag{4}$$

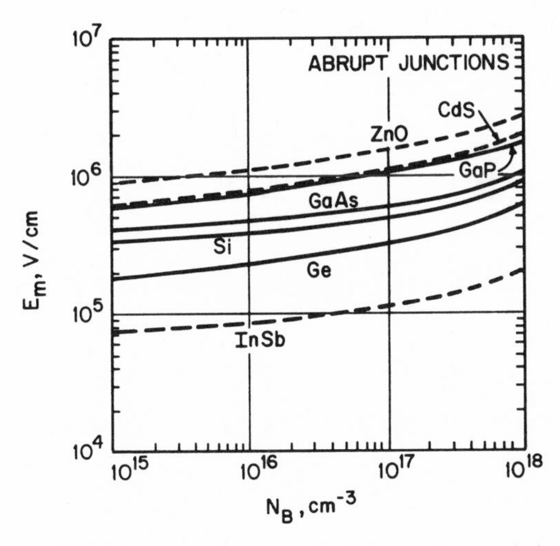

Figure V-2. Breakdown field vs carrier density for GaAs and Si.

Therefore, for a given cut-off frequency f_T, the maximum voltage V_m that can be applied to a device is proportional to the product of the breakdown field E_B and the average velocity υ_a. Both E_B and υ_a are related to the fundamental material parameters. For example, $V_m f_T \cong 2\text{x}10^{11}$ V/s for silicon because $E_B \cong 2\text{x}10^5$ V/cm and $\upsilon_a \cong 6\text{x}10^6$ cm/s; for GaAs in which $E_B \cong 3\text{x}10^5$ V/cm and $\upsilon_a = 1\text{x}10^7$ cm/s, $V_m f_T \cong 5\text{x}10^{11}$ V/s. Hence, the GaAs devices perform nearly twice as well as do comparable Si devices.

Equation (4) can also be interpreted as defining the upper limit on cut-off frequency. Since the minimum value of V_m must be the one that will supply a high electric field for the carrier to reach saturation velocity, say V, the cutoff frequency is proportional to $E_B\upsilon_a$, which is a semiconductor material property.

2. Current-Frequency Relationship

The maximum current of a semiconductor device is given by

$$I_m = \frac{Q}{\tau} \tag{5}$$

where Q is the total mobile charge in the device and τ is the transit time over the rf-modulation region. The maximum total mobile charge Q can be expressed as the product of the device output capacitance and the maximum voltage,

$$Q = CV_m \tag{6}$$

Substitution of equation (6) into (5) yields

$$I_m = \frac{CV_m}{\tau} \tag{7}$$

Combining equations (2), (3) and (7), we have

$$\frac{I_m}{C} = \frac{E_B L}{L\upsilon_a} = E_B\upsilon_a \tag{8}$$

Therefore, the current-to-capacitance ratio is related only to the material property $E_B\upsilon_a$ and is independent of device parameters such as doping density and area.

If we convert the capacitance C to the capacitive reactance X_C at the cut-off frequency f_T,

$$C = \frac{1}{2\pi f_T X_C} \tag{9}$$

Equation (8) becomes

$$(I_m X_C) f_T = \frac{E_B \upsilon_s}{2\pi} \tag{10}$$

where υ_s is the saturation velocity.

Thus, when the cut-off frequency f_T increases, the current-impedance product $I_m X_C$ needs to be proportionally decreased because the right-hand side of equation (10) is a material property that is roughly fixed for a given semiconductor material.

3. Power-Frequency Relationship

The maximum rf output power a given semiconductor device can generate can be determined by combining equations (4) and (10):

$$(V_m I_m) X_C f_T^2 = \left(\frac{E_B \upsilon_s}{2\pi} \right)^2$$

or (11)

$$X_C (P_m f_T^2) = \left(\frac{E_B \upsilon_s}{2\pi} \right)^2$$

This is the widely accepted equation stating that the power-frequency square-impedance product for a given semiconductor device is constant. In order to match the device impedance X_C to the conventional 50-Ω load for a reasonable operating frequency bandwidth and matching circuit loss, the value of X_C must assume some minimum value of say, 1 Ω. The above discussions lead to E. O. Johnson's Pf^2 law, which is widely accepted. It is customary to have Pf^2 = constant line included in a state-of-the-art output power figure. As an example, figure V-3 shows the best GaAs FET output power as a function of frequency as of February 1982. The best Pf^2 is 1000 W-GHz achieved by Mitsubishi. For the sake of completeness, we also include the state-of-the-art performance of the GaAs low-noise FET (fig. V-4), and of Gunn and IMPATT diodes (fig. V-5).

B. THERMAL LIMITATION

High-temperature operation not only degrades the device performance but also reduces the device life time. In the majority-carrier devices such as FETs, Gunn diodes, and IMPATTs, the carrier mobility and the saturation velocity decrease at high temperature. Therefore, the device performance, as described by equation (11), is degraded at high temperature.

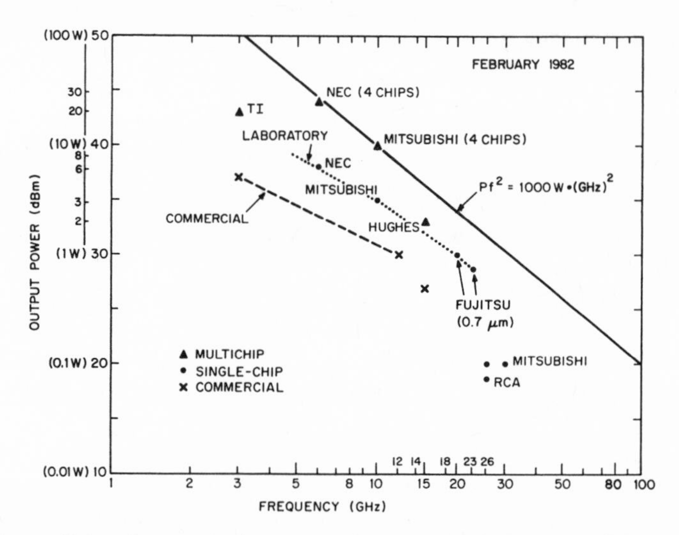

Figure V-3. State-of-the-art performance of GaAs power FET as of February 1982.

The rate of electromigration (EM) and diffusion also depends strongly on the operating temperature as depicted by the equation

$$t = t_o \exp\left(\frac{e\phi}{kT}\right) \tag{12}$$

or

$$\log_{10} \frac{t_1}{t_2} = 5.04 \times 10^3 \phi \left(\frac{1}{T_1} - \frac{1}{T_2}\right) \tag{13}$$

where t is the device life time, T is the absolute temperature in Kelvin, k is the Boltzman constant, and ϕ is the activation energy in electron volts. For the electromigration process, ϕ is of the order of 1 eV. In order to achieve long-lifetime operation, the maximum device operating temperature for space applications is about 120°C.

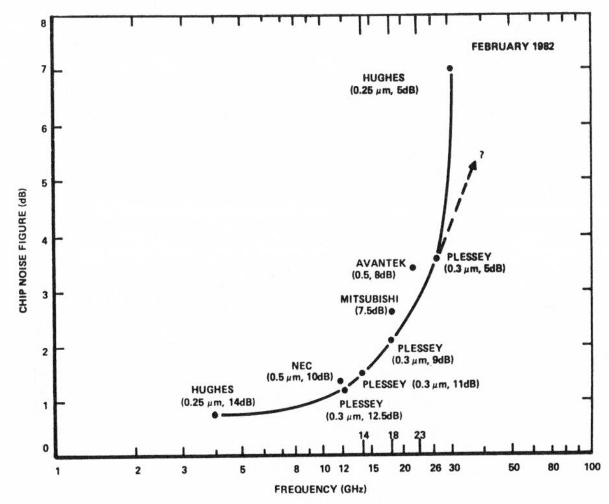

Figure V-4. State-of-the-art performance of GaAs low-noise FET as of February 1982.

For a given type of device, the maximum operating temperature depends strongly on the design and packaging technique of the device. We will show in this section that a properly designed microwave monolithic integrated circuit (MMIC) can have a maximum operating temperature below 120°C. Therefore, thermal consideration is, in general, not a limiting factor in the performance of a properly designed MMIC. We will use the thermal resistance of a GaAs FET as an example.

Figure V-6 shows the calculated thermal resistance of a GaAs FET as a function of the FET's gate-to-gate spacing, for various substrate thicknesses (C) and gate finger widths (W). The flip-chip mounting configuration is also included as dashed lines for completeness.

For a GaAs substrate thickness of 50 μm, the FET thermal resistance is less than 50°C-mm/W (fig. V-6). We can now estimate the actual channel temperature expected during rf operation of power GaAs FETs. A typical dc bias point yields 1-1.5 W/mm of dc input power (e.g., 100 mA/mm at 14 V). Depending on the rf power compression point, the net rf output power (i.e., output minus

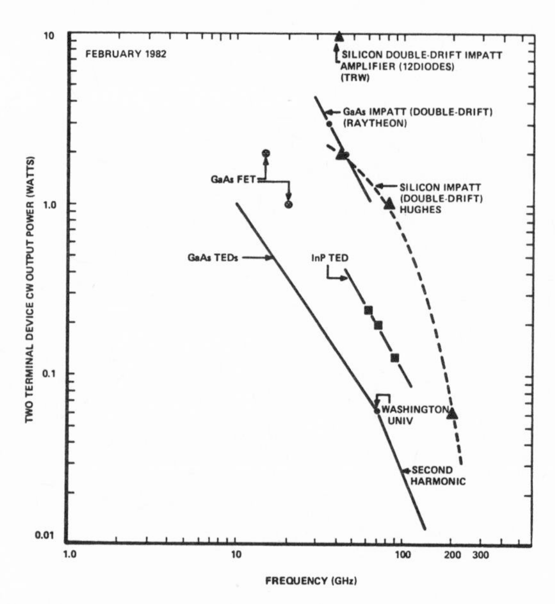

Figure V-5. State-of-the-art performance of Gunn (TED) and IMPATT devices as of February 1982.

input) usually falls in the 0.3- to 0.7-W/mm range so that the resulting dissipated power is about 1 W/mm or below. This means that, to a first approximation, the normalized thermal impedances shown in figure V-6 are close to the actual maximum temperature rise in the channel. The GaAs FET is therefore a remarkably cool device with channel temperatures less than 50°C above ambient, and an excellent active device for MMICs.

Detailed calculations and methods of measurement of GaAs FET thermal resistance can be found in reference 2.

C. PHYSICAL DIMENSIONS

In Section V.A (Electrical Limitation of Device Output Power at High Frequencies), we pointed out that the cut-off frequency of a semiconductor

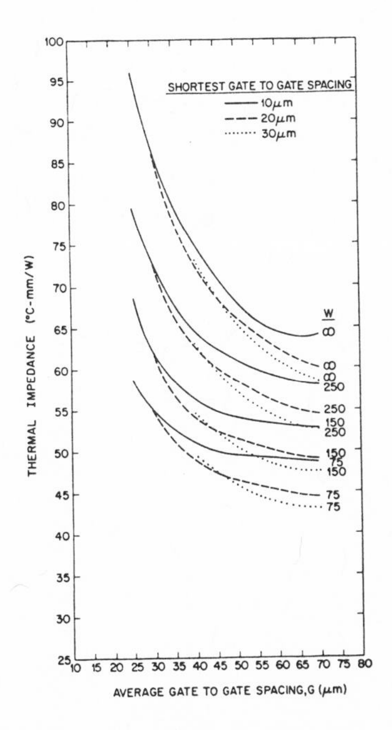

Figure V-6. Thermal impedance as a function of uniform gate-to-gate spacing (G) for various substrate thicknesses (C) and gate finger widths (W). Backside-mounted (solid lines) and flip-chip mounted (dashed lines) configurations are included. For these graphs, k = 0.038 W/°C·mm [2]. J. V. DiLorenzo and D. D. Khandelwal, Eds.: GaAs FET Principles and Technology. Artech House, Inc., 610 Washington Street, Dedham, Mass., 1982.

device is inversely proportional to the length of the rf-modulation region as given by the following equation [based on eqs. (1) and (2)]:

$$f_T = \frac{\upsilon_a}{2\pi L} \qquad (14)$$

where υ_a is the average carrier velocity and L is the length of the rf-modulation region.

Since one of the important devices considered for high-frequency MMIC implementation is the GaAs FET, we will calculate the projected cut-off frequency of the GaAs FET to show that the physical dimension is not a limiting factor to the FET high-frequency performance.

In the GaAs FET, rf modulation occurs under the gate. Therefore, the length of the rf-modulation region is roughly the length of the gate, ℓ_g. For our purposes, it is reasonable to assume that a gate length as short as 0.1 μm can be realized in a small-scale production environment. The average carrier velocity under a submicrometer gate length is of the order of 1.5 x 10^7 cm/s. Therefore, according to equation (14), the cut-off frequency of an FET with a 0.1-μm gate length is about 240 GHz, a frequency much higher than required for satellite communications in the foreseeable future. We conclude that the physical dimension will not be a limiting factor to the FET's high-frequency performance.

A submicrometer gate of the order of 0.1 μm can be fabricated by e-beam lithography, x-ray lithography, ion-beam lithography, and angle evaporation. When the gate length is 0.5 μm or smaller, the electrons do not have time to reach the heavy-mass, low-velocity state. The dynamic saturation velocity in this case is in excess of 2 x 10^7 cm/s, leading to an average velocity of about 1.5 x 10^7 cm/s.

To realize the potential performance of the high-frequency FET with very small gate length, the gate resistance and the rf phase equalization among the multiple gates have to be optimized. The technology of achieving low gate resistance and small phase unbalance for high-frequency MMICs requires further development.

D. DIGITAL MMICs

We have thus far considered mainly the limitations on analog (linear) MMICs. In this section, we will discuss the limitations on digital MMICs. Compared to the silicon digital IC, the GaAs MMIC has the advantages of high speed and/or low power dissipation.

In Section V.C, we have shown that the cut-off frequency of an FET with 0.1-μm gate length is about 240 GHz. This is equivalent to an ultimate intrinsic switching speed of about 7 x 10^{-13} s, or 0.7 ps. This intrinsic speed is again much higher than needed for satellite communications in the foreseeable future. We conclude that the limitation to digital MMIC performance is not the intrinsic switching speed. For future satellite communications, the limitation to digital MMICs is most likely the development of a logic technology that simultaneously meets the requirements of (a) maximum speed, (b) minimum dissipation, and (c) ease of implementation. Thus, it is necessary to consider carefully the complex tradeoffs between the various logic approaches and determine the optimum compromise.

There are two major logic approaches - depletion mode (d-mode), or normally-ON, and enhancement mode (e-mode), or normally-OFF, FET logic. The basic logic gates of these two types are shown in figure V-7. Propagation delay, power dissipation, and versatility of these logic types are compared in table V-1. In this comparison, we used the same currents for both types of devices, as the device currents determine the propagation delay and the rise and fall times. In determining the best performance possible from each device, the effects due to parasitics (e.g., series resistances, stray capacitances, etc.) were not included. In the actual realization such effects cannot be separated and hence the performance will be below these expectations.

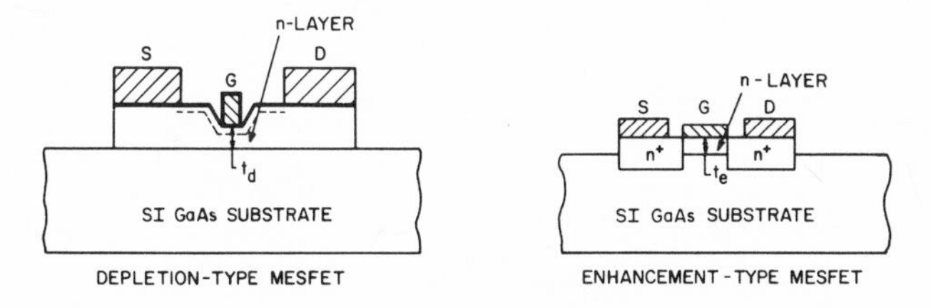

Figure V-7. d- and e-mode MESFETs and logic gates.

Table V-1 makes it clear that the e-mode logic has the performance advantage of higher speed and lower power dissipation but that it also has less noise margin and requires better material and processing uniformity than the d-mode FET logic.

To further illustrate the tradeoffs between e-mode and d-mode logic, we will consider the ring oscillator circuit. Ring oscillator circuits can be used to determine the propagation delay and power dissipation per logic gate. Ring oscillator results do not provide any insight into noise margins, tolerances, and signal levels. Often, the loading (i.e., fan-in and fan-out) varies from 1 to 3. Therefore, ring oscillator results cannot be used to predict system performance. However, ring oscillator results can be used to test device models and compare various technologies. A summary of the performance

Table V-1. COMPARISON OF e-MODE AND d-MODE LOGIC

Parameters	Enhancement Mode	Depletion Mode	Figure of Merit
Delay	$\tau d = \sqrt{3}\ t_r$	$\tau d = 3\ t_r$	$\tau_{dd}/\tau_{de} = \sqrt{3}$
Pinchoff Voltage (V_p)	V_B	$4\ V_B$	
Variation in V_p	$\Delta V_p = 0.1$ V	$\Delta V_p \geq 0.5$ V	
Power Dissipation	.... $P = \frac{\alpha}{2} V_p$ $\alpha = 1.5\text{-}2.0$	$2\varepsilon n e V_B \cdot \upsilon_s W$ .... $\alpha = 4.6$	$P_{dd}/P_{de} = 12$
Active-Layer Thickness	900-1000 Å	1800-2000 Å	
Load Requirement	Ungated FET or resistive load; I_L/I_s not well defined	FET with $V_{GS} = 0$ for load; I_L/I_s well defined	
Logic Voltage Range	$V_H = 1.0$ V; noise margin for $V_L \simeq 100\text{-}200$ mV	$V_L = -2.0$ V, $V_H = 0.5$ V; noise margin >0.5 V	
Fan-in/Fan-out Limits	Fan-in limited by leakage currents, fan-out limited by load current	Fan-in and fan-out not limited by devices; limited only by layout or by response time desired	

Symbols:

τ_d = propagation delay; t_r = transit time

V_p = channel pinchoff voltage

V_L = logic low; V_H = logic high

υ_s = saturation velocity; W = channel width

of ring oscillators reported by various industry leaders in GaAs IC technology is given in tables V-2 and V-3. It is evident from these two tables that the enhancement-mode logic gates are as fast, if not faster, than depletion-mode logic gates. Also, the power dissipation per gate is lower by at least a factor of 2 for e-mode logic.

Table V-2. PERFORMANCE OF 1.0-μm GATE-LENGTH d-MODE MESFETs IN RING OSCILLATORS

Name of the Organization	Propagation Delay (ps)	Power Dissipation per Gate (mW)	Fan-out	Type of Tech-nology
Hewlett-Packard	60	10	1	BFL*
Rockwell-International	75-110	1.0-2.2	1	SDFL**
Thomson-CSF	130	40	2	BFL
RCA Laboratories	125	13.5	2	BFL
Hughes Aircraft	83	5.6	1	BFL

*Buffered FET logic.
**Schottky-diode FET logic.

Table V-3. PERFORMANCE OF 1.0-μm GATE-LENGTH e-MODE MESFETs IN RING OSCILLATORS

Name of the Organization	Propagation Delay (ps)	Power Dissipation per Gate (mW)	Fan-out
Nippon Telegraph & Telephone Public Corp.	45 39.5	0.82 4.05	1 1
Fujitsu	50	5.7	1
Bell Laboratories	30	--	1

Further technology development in material and process uniformity is therefore essential to realizing the full potential of enhancement-mode logic.

E. RECOMMENDATIONS

The physical limitations discussed above indicate that there are several technology areas that will affect the limits. These areas include material, lithography, processing technology, and new device structure, as described below.

- Material. The ideal material characteristics are high carrier mobility, high saturation velocity, high breakdown field, and high thermal conductivity. While it is difficult to achieve all of the above properties simultaneously, there are a number of materials that exhibit promising characteristics. For example, modulation-doped GaAlAs-GaAs and GaInAs have demonstrated mobilities higher than that of GaAs, while silicon carbide has high thermal conductivity and high breakdown fields together with reasonable carrier mobility and saturation velocity. Further material research such as MBE technology will improve the fundamental device performance limit by improving the material figure of merit, $E_B \upsilon_s$.

• Lithography. To achieve high yield and high performance in the high-frequency MMICs, the technology for achieving submicrometer gate length with good dimensional uniformity and low gate resistance needs to be developed and refined. Ion beam, x-ray and e-beam lithography are the technologies that will lead to submicrometer pattern generation.

• Processing Technology. It is important to improve and refine the MMIC process technology to achieve high device yield. This technology includes, for example, air bridges, bathtub via holes, low-loss dielectric materials, smooth conductor edges, high-density metallization, and device packaging.

• New Device Concepts. As discussed earlier, the material figure of merit, $E_B \upsilon_s$, is the ultimate limit to device performance. The actual device performance, however, is much below this limit. In general, different device operating principles and device designs will lead to different performance limits. It is therefore highly beneficial to continue searching for new device concepts and new device structures. Some examples are the solid-state traveling wave FET and the vertical FET. With continuous research and development of new devices, the performances of high-frequency MMICs will experience not only an evolutionary improvement over the present state of the art, but may also include revolutionary breakthroughs.

F. REFERENCES

1. Johnson, E. O.: Physical Limitations on Frequency and Power Parameters of Transistors. RCA Rev., vol. XXVI, no. 2, June 1965, pp. 163-177.
2. Wemple, S. H., and Huang, H. C.: Thermal Design of Power GaAs FETs; in DiLorenzo, J. V., and Khandelwal, D. D., eds.: GaAs FET Principles and Technology. Artech House, Inc., Dedham, Mass., 1982, pp. 313-347.

VI

Task 5: Problems in Implementation

In this section we examine selected problems encountered, or likely to be encountered, when advanced techniques involving III-V compounds are used in components for space communications systems. First, the experiences in building the solid-state power amplifier (SSPA) for RCA's SATCOM - the first such unit designed for a satellite communications transponder - are reviewed. Next, problems in implementing monolithic (or miniature hybrid) circuits, linear amplifiers, and gigabit-rate digital circuits are discussed. Finally, device reliability considerations are reviewed - again based on experiences with present SSPAs.

A. EXPERIENCE WITH THE RCA SSPA

The first solid-state power amplifier (SSPA) designed for satellite transponder service was designed and space-qualified by RCA and is now in production at RCA's Astro-Electronics facility.

RCA has acquired considerable background and experience in the design, test, flight qualification, and production of GaAs FET amplifiers to replace traveling-wave tubes for satellite downlink communications systems. The unit provides a power output of 8.5 W, a gain of 60 dB, and a typical efficiency of 35% over the 3.7- to 4.2-GHz band. This amplifier, which consists of six sets of cascaded stages, as shown in figure VI-1, is built by the use of microwave integrated-circuit (MIC) technology, with each amplifier module designed to permit stand-alone testing and pretuning before being integrated to higher assembly levels.

The experience and background accumulated during the electrical and mechanical design phases, productization, and qualification programs for the 8.5-W SSPA have since been applied to similar programs for applications at X-, Ku-, and K-band frequencies.

Some of the problems and related solutions for implementing the MIC technology used in the aforementioned satellite amplifier are as follows:

(1) In soldering/bonding and related operations, caution must be exercised to ensure the minimization of gold-circuit leaching when components are soldered into place on gold-printed circuit substrates. This leaching, which is a function of time and temperature, erodes the gold and can lead to circuit

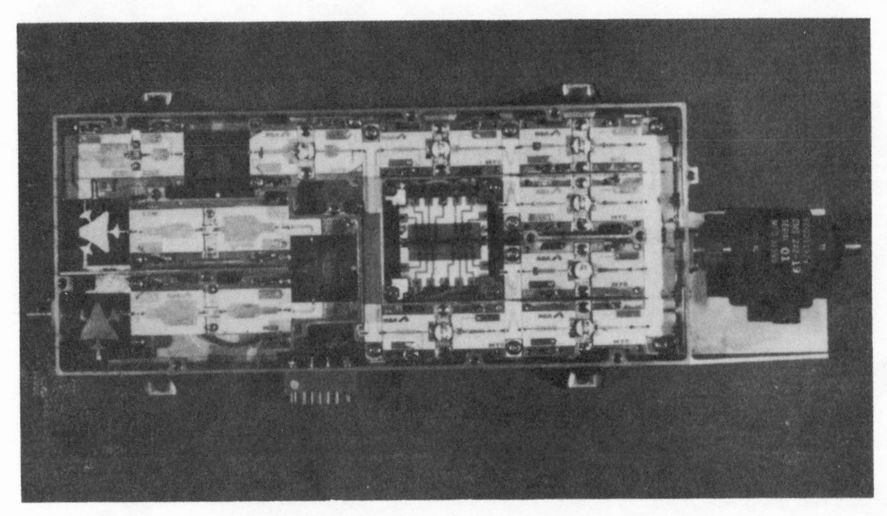

Figure VI-1. 8.5-W C-band microwave-IC SSPA.

failure in areas of high current density. For optimum reliability, soldering should be avoided wherever possible. Parallel-gap welding or thermal-compression bonding should be used wherever possible.

Corrosive soldering or flux with strong activators should be avoided if possible. Corrosive residue is difficult to completely remove with conventional electronic wash and rinsing processes, and residues can lead to erosion and premature failure with life.

(2) When multiple passive and active components are soldered on the same substrate, it may be necessary to use soldering alloys of varying flow temperature to minimize reflow and the movement of critically placed components.

(3) A comprehensive quality assurance program must be implemented. All incoming components, active/passive, must be properly and sufficiently specified and tested to ensure compliance with rigid procurement specifications. This will enhance uniformity and reproducibility. All components must undergo qualification acceptance tests; processes and manufacturing methods must be well documented, standardized, and controlled.

(4) All components not space-qualified on related programs must undergo sufficient testing to ensure compliance with mechanical, thermal, electrical, and environmental tests.

(5) Substrates must be tested to ensure meeting the proper plating adherence to avoid circuit lift-off.

(6) Test fixtures for rf functional circuits, modules, and higher levels of integration must be designed without compromising interface impedances or line length so that erroneous data are not obtained. This is particularly important where parasitic tuning is used to provide for enhanced bandwidth or gain flatness. Tuning a circuit with chip capacitors to selectively correct for gain vs frequency fall-off, caused specifically by the test fixture, may result in an intolerable gain contour when the test fixture is removed.

(7) Ground plane discontinuities must be avoided, specifically in the area where adjacent amplifier modules are interconnected with rf output and input lines. The ground plane under the microstrip circuit line must be continuous to obtain the proper circuit impedance to avoid reflections and gain perturbations.

(8) Periodic ground via lines or septa should be used to minimize circuit inductance in the ground return so as to reduce the limits on the frequency bandwidth and on the high-frequency gain.

(9) Hermetically sealed active devices are preferred, so as to minimize mechanical packaging design. Active chip devices for higher frequency applications may require hermetically sealed packages for the amplifier to meet environmental humidity requirements.

(10) Bias feed lines must be adequately filtered to eliminate feedback, spurs, and instabilities. Quarter-wave line transformers used to shunt-feed the drain line in an FET power circuit must simultaneously sustain high current densities without undue voltage drop and present a high impedance to rf-conducted leakage. These are conflicting requirements.

(11) Adequate Filtercon insulators (made by Erie Technological Products, Inc., Erie, Pa.) should be used in bias feed lines to increase isolation between stages and eliminate bias oscillation and instabilities.

(12) Partitioning between stages or effective isolation by means of waveguide-below-cutoff techniques to inhibit random rf coupling and feedthrough is required. This will help eliminate fine-grain gain variations and rf stability problems.

(13) Differential thermal expansion between circuit substrate and support package must be minimized. This is especially important under repeated thermal cycling where stresses causing rupture can occur. In addition, ground plane discontinuities can result, causing rf reflections as well as gain and power perturbations.

B. MONOLITHIC AND MINIATURE HYBRID CIRCUITS

The relative advantages and disadvantages of monolithic vs miniature hybrid circuits were discussed in some detail in Section IV (Task 3). From a practical point of view, both of these types of circuits will have to be made adaptable to automated testing and trimming adjustments before they can be used in the quantities needed for implementing, for example, large phased-array antennas.

The driving parameters for the use of monolithic microwave circuits and/or miniature hybrids are uniformity and low cost. Even though present-day GaAs FETs are proving cost effective for "normal" satellite transponder service, their use would be prohibitively expensive in a large antenna system. At present, an attempt is normally made to extract maximum performance from every FET stage (because of its cost), which usually involves characterizing each individual transistor and optimizing its matching circuit (by computer-aided techniques).

In the future, as large quantities of low-cost circuits come into use, the philosophy will be the opposite: to use the active devices below their optimum capabilities, so that small variations in device and circuit characteristics will not result in nonuniform performance of the monolithic or miniature hybrid modules.

Thus, before large quantities of high-performance, low-cost monolithic/ miniature hybrid modules can be manufactured, extensive effort will be required to

- desensitize circuits from overdependence on device characteristics;
- develop automatic methods for testing monolithic/miniature hybrid circuits;
- develop methods for the automatic trimming of circuits; and
- make circuit fabrication techniques compatible with the emerging robotics approaches.

C. $Ga_{0.47}In_{0.53}As$ MISFETs FOR LINEAR MICROWAVE AMPLIFIERS

To overcome limitations in the high-frequency operation of GaAs FETs, ternary compounds are being investigated at RCA Laboratories and elsewhere. As explained below, metal-insulator-semiconductor field-effect transistors (MISFETs) made of ternary compounds appear to provide better linearity than GaAs metal-semiconductor field-effect transistors (MESFETs). This may have great potential for future space communications programs.

The investigations center around $Ga_{0.47}In_{0.53}As$ lattice matched to semi-insulating InP as the semiconductor material for such applications. Theoretical and experimental studies have shown that this material has the potential advantages of higher low-field electron mobility and peak electron drift velocity over GaAs [1-3]. However, the bandgap of this material is quite low (0.72-0.75 eV), resulting in a Schottky-barrier height for typical metals of only 0.3 V [4]. This low Schottky-barrier height presents problems in fabricating Schottky-barrier-gate FETs. Various approaches to circumvent this problem, including the use of p^+n junction gates [5] and heterojunction gates [6] are being investigated in many laboratories. At RCA, we have taken the approach of developing MISFETs [7].

In this approach, the Schottky-barrier gate is replaced by a metal-insulator gate, making the gate purely capacitive, and eliminating the problems of low Schottky-barrier heights.

There is a potential advantage to using an insulated gate. As there is no Schottky diode, the voltage swing on the gate is limited only by the breakdown of the insulator dielectric, unlike the case of the Schottky gate, where the voltage swing is limited by the forward voltage of the Schottky diode in the positive direction, and by the reverse breakdown of the Schottky diode in the negative direction.

Figure VI-2 shows the I-V characteristics of a depletion-mode (normally-on) $Ga_{0.47}In_{0.53}As$ MISFET. The left-hand trace shows the characteristics for positive gate bias, and the left-hand trace for negative gate bias. It can be seen that with positive gate bias there is a considerable increase in the drain current. Thus, the device can be operated at zero bias, and as the gate voltage swings about zero, the drain current is increased or decreased. This makes simple bias circuits possible, and the potential for high efficiency and linearity exists.

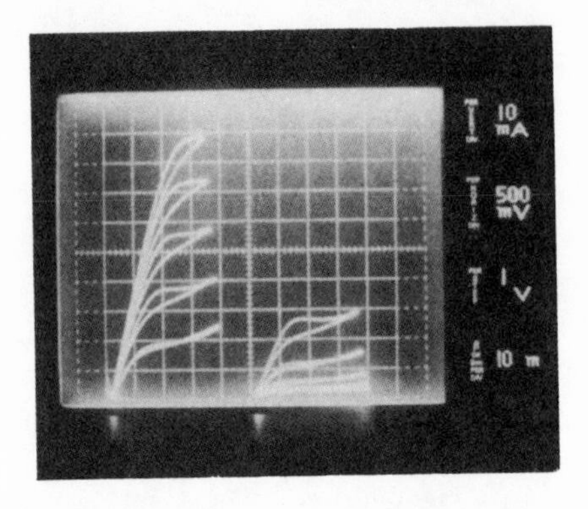

Figure VI-2. I-V characteristics of depletion-mode MISFET.

Our results so far have been very encouraging. Table VI-1 lists results obtained on a ternary depletion-mode MISFET with a 1.5-μm channel length and 600-μm channel width, operated at a drain voltage of 4.5 V at various frequencies.

TABLE VI-1. $Ga_{0.47}In_{0.53}As$ MISFET rf PERFORMANCE

Wafer Q691 L_g = 1.5 μm W ∼ 600 μm $C_{ox} = 3.73 \times 10^{-8}$ F/cm^2 V_{DS} = 4.5 V

Frequency (GHz)	Gain (dB)	Power Output (mW)	Power-Added Efficiency (%)
4	8.5	111.7	36.0
6	3.8	119	30.6
11	3.0	90	22
12	3.0	59	9.4

The relatively low power output per millimeter of channel width (0.15 W compared with ∼0.5 W for GaAs MESFETs) is partly due to the low drain voltage, and, for this device, a lower-than-optimum zero-bias drain current. We believe that a significant increase is possible. Nonetheless, the efficiencies and gains obtained are very encouraging.

Recent measurements have shown that depletion-mode ternary MISFETs also exhibit good linearity. For a device similar to that shown in table VI-1, operated at a drain voltage of 4.1 V and zero gate bias, measured at 4 GHz with two tones separated by 10 MHz, we have obtained third-order intermodulation distortion (IMD) 25.5 dB down, with a power output of 109 mW, gain of 9.4 dB, and power-added efficiency of 32.7% at the 1-dB gain comparison point.

This combination of low third-order IMD and high power-added efficiency makes the MISFET a very attractive candidate for linear microwave power amplifiers.

No attempt was made to optimize either the device or the measurement conditions to obtain low IMD. It is not fully understood whether or not the MISFET is inherently more linear or more efficient than a MESFET. However, the results are sufficiently good to warrant further investigation of the potential of these devices for linear microwave power amplifiers.

D. HIGH-SPEED DIGITAL CIRCUITS

Technology developments for the fabrication of GaAs digital monolithic integrated circuits have been reported by several research laboratories from all over the world. GaAs ICs are consistently faster than their silicon counterparts. GaAs technology is still very young and optimum tradeoffs between normally-on (or depletion mode) and normally-off (or enhancement mode) logic circuits are being studied. Impressive performance is reported with depletion-mode [8] as well as enhancement-mode FET logic [9]. However, several problems remain to be solved at this time before GaAs digital ICs can be produced consistently with reasonable yields and still meet the stringent specifications. The problem areas to be addressed are (a) materials, (b) processing, and (c) packaging and measurements.

1. Material Requirements

GaAs digital ICs require thin active layers (900-1500 Å) on semi-insulating substrates. The drift mobility for electrons in such thin layers should be maintained as high as possible. For optimum device performance, several different types of doping profiles have to be generated selectively in relatively small geometries. Of course, interface (active-layer/substrate) trapping problems should be eliminated or minimized. Also, the doping density and thickness of the active layer must be controlled to within 5% for MSI/LSI circuits. These problems pose a formidable task in obtaining starting material for GaAs IC work. We believe that such stringent material requirements can be met only by ion implantation or molecular-beam epitaxy.

a. Ion Implantation

Ion implantation is a very powerful technique for obtaining uniform, controlled, and selective doping of semiconductors. Ion implantation in Si for LSI/VLSI work is a widely used technique and is rapidly becoming a widely used tool in GaAs work. Direct implants into SI GaAs are essential for GaAs multi-gigabit-rate digital ICs. The quality of the implanted layers is measured by factors such as the electron mobilities, doping profile, activation efficiency, pinchoff voltage control, and reproducibility. All these parameters are strongly influenced by the substrate quality [10] and the annealing method. Undoped SI GaAs substrates proved to be better candidates for ion implantations. Substrate vendors and research laboratories are using the Melbourn liquid encapsulated Czochralski (LEC) equipment for producing SI GaAs substrates. Large-diameter (8 cm) substrates were grown by this method. Some research laboratories claim (e.g., Rockwell International) that these substrates are superior to those grown by other methods. However, commercially available LEC substrates are not of as high a quality as that reported by research laboratories. Therefore, it is still necessary to qualify SI GaAs substrates from each boule and each vendor. There are some guidelines for qualification tests, but each laboratory uses its own criteria for accepting the substrates for ion implantation. High-resistivity buffer layers grown on the substrates may reduce some of the stringent requirements for SI GaAs substrates. However, very few laboratories have the capability to grow such buffer layers reproducibly. Hewlett-Packard Company is one of the few companies that uses high-resistivity buffer layers grown by liquid-phase epitaxy (LPE) for ion implantations.

The method of annealing implantation damage and activating the implanted ions is critical to achieving high-quality, thin active layers on SI GaAs substrates. Impurity redistribution and diffusion are characteristic of the annealing method employed. Laser annealing [11], electron-beam annealing [12], and annealing with radiation from a quartz-halogen lamp may prove to be better than thermal annealing. However, these techniques are not yet developed to the extent that they can be applied to actual device fabrication.

b. Molecular-Beam Epitaxy (MBE)

MBE is an epitaxial deposition process using molecular or atomic beams in an ultrahigh vacuum ($\sim 10^{-10}$ Torr). The substrate temperature is kept at 500-600°C. The growth rate is very slow (~1 μm per hour), making it possible to grow ultrathin layers (~50Å). Complex doping profiles and sophisticated device structures can be grown by successive depositions without thermal diffusion or redistribution. The growth is determined by complex kinetic reactions at the substrate, and stoichiometric films may not result even for congruent molecular beams. Incorporation of dopants into semiconductors is a complex process. Several laboratories reported high-quality MBE GaAs growth techniques and device performance. At best, the performance of MBE-grown devices approached that of the best epitaxially grown (by vapor phase) devices. Therefore, the potential is there for MBE. The superiority of MBE is established in the growth of multi-semiconductor structures such as AlGaAs/GaAs. However, MBE has not yet been developed into a production process.

2. Processing of GaAs ICs

GaAs ICs consist of MESFETs and Schottky diodes for depletion-mode logic, and MESFETs and load resistors (linear/nonlinear) for enhancement-mode logic. Technology development has been reported by several research laboratories for d- or e-FET logic. In spite of the limited success by various laboratories, it is not possible to implement GaAs LSI circuits reproducibly. Ohmic-contact and Schottky-gate technologies are adequate but need further refining. An evaporated Au:Ge/Ni/Au metal system with different thicknesses, compositions, and heat treatment is employed for ohmic contacts. Even though several studies have reported optimum parameters, occasional bubbling of the contacts was observed. Severe bubbling of the contacts limits photoresist patterning for subsequent processing steps (particularly for 1.0-μm gates). Refractory metals are generally used for Schottky gates, the exception being aluminum. Such Schottky barriers are stable up to 350-450°C. In the self-aligned gate structure proposed by Nippon Telegraph & Telephone Public Corp. (NTT) [13], the gates must be stable up to 850°C. Ti:W/Si [14] or Mo [15] Schottky gates may withstand such high temperatures. These technologies are at a very preliminary stage and require further studies.

To decrease the propagation delay and increase the speed performance of MESFET logic devices, work is progressing on two different fronts: (i) to minimize the parasitics (particularly the gate-source and gate-drain series resistances) and (ii) to use devices with smaller geometries (submicrometer gate lengths).

a. Minimizing the Series Resistance

Reduction of series resistance is mainly achieved by the use of heavily doped, deep n^+-regions under the source and drain contacts. This can be easily achieved by selective ion implantations. However, a laser beam or electron beam or radiation from a halogen-quartz lamp must be employed for annealing to avoid lateral diffusion (or spreading) of the dopant. In addition to using n^+-ohmic contacts, self-aligned gate structures may yield the optimum devices. These technologies are at an early stage of development, and further studies are required before they can be employed for LSI circuit fabrication.

b. Reduction of the Device Geometries

The load capacitance in MESFET logic is the gate-source capacitance of the Schottky gate of the MESFET. Any reduction in gate length will result in a corresponding reduction of gate capacitance. Therefore, submicrometer gate-length devices are preferred for logic circuits. Such submicron gate devices can only be fabricated by electron-beam or x-ray lithography. Electron-beam lithography [16] has been in use for saveral years; its main drawback is the throughput. Machines with higher beam currents are being developed and should be available within a few years. Of course, these machines cost several million dollars ($3-6 million). X-ray lithography is also at an early stage of development. Recently, 0.3-μm gate-length devices [17] have been fabricated by x-ray lithography in Si. The types of masks required and the photoresists used have not yet been optimized. Further studies are needed to optimize these factors.

c. Process Yield

The fabrication of GaAs MSI/LSI circuits on a reproducible basis requires a high process yield. Here, process yield is used in a general sense so as to include device parameters. Device pinchoff voltages and currents are very important. Pinchoff voltage determines the available noise margins, and the device currents determine the speed of the circuits. At most a 5-10% variation in pinchoff voltage and device currents across the wafer is desirable for good process yield. Unless further improvements are made in materials and device technologies, GaAs digital ICs cannot be implemented.

3. Packaging and Testing

MESFET logic-gate rise, fall, and propagation delay times are a function of fan-out and, consequently, capacitive loading. This problem is similar to that experienced with emitter-coupled logic (ECL) devices. Therefore, when FET logic circuits are operating near their maximum speed limit, fan-out should be restricted. Signal lines should be kept as short as possible to minimize ringing and overshoot, as well as to simplify timing considerations arising from the propagation delay of the signal along a conductor. Intrinsic inductance and capacitance are reduced by shortening the line. Any signal path may be considered a form of transmission line. If the propagation delay along the signal path is short compared with the rise time of the signal, any reflections are masked in the rise time and are not seen as overshoot or ringing. The maximum open line length may be given by

$$\ell_{max} \leq \frac{t_r}{2t_{pd}}$$

where t_r = rise time

t_{pd} = propagation delay of the line per unit length

= $1.017 \sqrt{\varepsilon_r}$ (ns/ft)

where ε_r is the relative dielectric constant. On GaAs substrates ($\varepsilon_r = 12$), for 100-ps rise-time circuits, the maximum length of the open line is about 160 mil or 0.4 cm.

In the MESFET logic, the Schottky gates of the individual devices are operated in the reverse-bias region (for d-FETs) or weakly forward conducting region (for e-FETs). As a result, the load is mostly capacitive. This lets one design logic gates with small geometry (i.e., low current) devices, even though the rise and fall times are of the order of 100 ps. Such logic gates cannot drive 50-Ω lines. Special interface circuits (multiple-stage source-follower circuits) may be required to drive 50-Ω lines. The dc power dissipation in these circuits will be very high, and it is not clear whether these circuits should be included on or external to the IC.

Digital ICs have several input and output ports. Interface circuits have to be developed with the chip area and the dc power economically utilized. The layouts of the signal lines determine the crosstalk. In the presence of crosstalk, noise margins are reduced. So far, no results on crosstalk in GaAs ICs have been reported.

Test equipment available for high frequencies or fast rise times is generally of the 50-Ω characteristic impedance type. This implies that the input signal lines have to be terminated in 50 Ω, and the output signal lines must be capable of driving 50-Ω lines. Also, test equipment, such as a pulse generator or word generator, is not available above 350 MHz. Furthermore, MESFET logic circuits require a 1- to 2-V swing for the logic. For high-speed testing, the equipment is designed with ECL in mind. The logic swing for ECL is only 0.8 V. This poses a serious problem. Test equipment and test procedures for high-speed GaAs ICs have to be designed.

E. RELIABILITY CONSIDERATONS

The continuing development of technology in the semiconductor industry will bring with it increased difficulties in assuring highly reliable products for space-flight missions. Many of the classical failure mechanisms associated with semiconductor devices that are now well managed and controllable will reappear as an increase in packing densities and a shrinkage in linewidths to submicrometer size. Additionally, it is likely that hitherto unforeseen failure mechanisms will appear as dimensions shrink and densities increase.

The increasing utilization of active surface area of semiconductor devices places severe operating constraints on these devices from a reliability viewpoint. The principal concern must be the removal of heat from the device area. It is well known that most semiconductor failure mechanisms are accelerated by increasing temperatures. In some cases, a temperature difference of +10°C can halve the lifetime of these devices. Very high packing densities can increase device operating temperatures by several tens of degrees even when special care is paid to thermal design of the device and mounting schemes. This consideration will eventually force a tradeoff between device performance and lifetime.

As device size and linewidths shrink, electromigration will once again limit the operating lifetimes for these devices. With increasing device complexity, current-carrying lines will be forced to support larger current densities at higher operating temperatures. Device performance constraints, such as speed and noise immunity, will limit the minimum voltage and current-operating parameters. As line dimensions shrink by one half, densities will increase by a factor of 4. It is then very easy to pass from safe operating current densities into regions where lifetime is considerably reduced.

Increasingly, as device dimensions shrink, random catastrophic failures will become important in determining device life. Hillock growth, now seen in many semiconductor devices, will have a greater probability of shorting to a critical area of the device. This result is unpredictable except in a random sense, and has the possibility of catastrophically failing the device. Crystal dislocations will also play a critical role in determining not only device yield but also device reliability. These crystal imperfections will have the potential of being a significant fraction of device and linewidth size. Thus reliability problems associated with their long-term growth will have a much greater impact than they have at present, when their dimensions are small compared with device or line dimensions. Both of these problems (hillock and dislocation growth) are relatively immune to device screening.

The problems discussed to this point are all associated with extending present technology to its extreme limits. This technology is essentially two-dimensional, with most devices and interconnects formed near the surface of the chip. Recent work in epitaxial layer formation suggests that it may be possible to form three-dimensional semiconductor chips, with devices formed in the body of the material as well as near the surface. A rudimentary example of this type of technology is the power MOS transistor, in which the drain current of thousands of unit cells is collected in the body of the silicon chip, and drain contact is made to the back of the wafer.

This type of multilevel technology will present new problems in reliability and failure analysis. Determination of specific device operating temperatures will then require sophisticated thermal analysis. Removing heat from the devices will continue to be a complex problem. New chip design guidelines will be required to maximize thermal flow. Failure analysis of parts will be extremely difficult, requiring careful and tedious processes to carefully remove device layers.

As current technology is pushed to its limits and new technology is developed, careful attention must be given to performance/reliability trade-offs. Current technological trends point to short-term reliability loss. This will eventually be overcome by process engineers improving or changing those basic components of a semiconductor device that are present reliability risks. If this proves difficult, increased reliability will be gained by designing for maximum device performance and device standby redundancy to increase overall system life.

F. REFERENCES

1. Littlejohn, M. A., Hauser, J. R., and Gibson, T. H.: Velocity-Field Characteristics of $Ga_xIn_{1-x}As_yP_{1-y}$ Quarternary Alloys. Appl. Phys. Lett., vol. 30, no. 5, 1977, p. 242.
2. Greene, P. D., Wheeler, S. A., Adams, A. R., El-Sathby, A. N., and Ahmed, C. N.: Background Carrier Concentration and Electron Mobility in LPE-Grown $Ga_xIn_{1-x}As_yP_{1-y}$ Layers. Appl. Phys. Lett., vol. 35, no. 1, 1979, p. 78.
3. Leheny, R. F., Ballman, A. A., DeWinter, J. C., Nahory, R. E., and Pollock, M. A.: Compositional Dependance of the Low-Field Mobility of $Ga_xIn_{1-x}As_yP_{1-y}$. J. Electron. Mater., vol. 9, no. 3, 1980, p. 561.
4. Kajiyana, K., Mizushima, Y., and Sakata, S.: Schottky Barrier Heights of η-$Ga_xIn_{1-x}As$ Diodes. Appl. Phys. Lett., vol. 23, 1973, p. 458.
5. Leheny, R. F., Nahory, R. E., Pollock, M. A., Ballman, A. P. Becke, E. D., DeWinter, J. C., and Martin, R. J.: An $In_{.53}Ga_{.47}As$ Junction Field-Effect Transistor. IEEE Electron Device Lett., vol. EDL-1, no. 6, 1980, p. 110.
6. Barnard, J., Ohno, H., Wood, C. E. C., and Eastman, L. F.: Double Heterostructure $Ga_{.47}As$ MISFETs with Submicron Gate. IEEE Electron Device Lett., vol. EDL-1, no. 9, 1980, p. 174.
7. Gardner, P. D., Narayan, S. Y., Colvin, S. and Yun, Y. H.: $Ga_{.47}In_{.53}As$ Metal Insulator Field-Effect Transistors (MISFETs) for Microwave Frequency Applications. RCA Rev., vol. 42, no. 4, Dec. 1981, pp. 542-556.
8. Liechti, C. A., et al.: A GaAs Word Generator Operating at 5 Gbits/s Data Rate. IEEE Trans. Microwave Theory Tech., vol. MMT-30, no. 7, July 1982, pp. 998-1006.
9. Ino, M., Hirayama, M., Ohwada, K., and Kurumada, K.,: GaAs/KB Static RAM with E/D MESFET DCFL. GaAs IC Symp., New Orleans, LA, Nov. 1982, Dig. Tech. Papers, pp. 533-538.
10. Hobgood, H. M., et al.: High Purity Semi-insulating GaAs Material for Monolithic Microwave Integrated Circuits. IEEE Trans. Electron Devices, vol. ED-28, no. 2, Feb. 1981, pp. 140-149.

11. Liu, S. G., et al.: Annealing of Ion-Implanted GaAs with a Pulsed Ruby Laser. Symp. Proc. on Laser and Electron Beam Processing of Materials, Academic Press, New York, 1980, pp. 341-346.
12. Pinnetta, P. A., et al.: Pulsed E-beam Ruby Laser Annealing of Ion--Implanted GaAs. Symp. Proc. on Laser and Electron Beam Processing of Materials, Academic Press, New York, 1980, pp. 328-333.
13. Yamasaki, K., et al.: Self-Aligned Implantation for n^+-Layer Technology (SAINT) for High-Speed GaAs ICs. Electron Lett., vol. 18, no. 3, Feb. 4, 1982, p. 119.
14. Yokoyama, N., et al.: TiW Silicide Gate Self-Aligned Technology for Ultra-High Speed GaAs MESFET LSI/VLSI. To be published in IEEE Trans. Electron Dev.
15. Lepselter, M. P.: Silicon Picocircuits. GaAs IC Symp., San Diego, CA, Oct. 27-29, 1981, Dig. Tech. Papers (Paper 6).
16. Greiling, P. T., et al.: Electron-beam Fabricated High-Speed Digital GaAs ICs. Proc. IEEE, vol. 70, no. 1, Jan. 1982, pp. 52-58.
17. Mukherjee, S. D., Palmstron, C. J., and Smith, J. G.: The thermal stability of thin layer transition and refractory metallization on GaAs. J. Vac. Sci. Technol., vol. 17, no. 5, Sept./Oct. 1980, pp. 904-910.

VII

Task 6: Recommendations for Implementation

A. SCOPE

In this section we present recommendations on how to take advantage of the technical capabilities of III-V compounds to enhance the performance of future space communications systems. It is important to emphasize that:

(a) The recommendations are for technology-development programs, as opposed to specific mission-related programs. The RCA SSPA development for the SATCOM system may serve as an illustration: The technology of GaAs power FETs operating at 4 GHz was considered "ready" for exploitation several years ago. It took considerable skill, tight planning, commitment of considerable funds, and several important ancillary programs - such as reliability proof, automated testing, "transfer of technology" to a manufacturing group, and the establishment of quality-assurance procedures - before this "ripe" technology could be utilized in spaceborne transponders.

(b) In estimating costs, it was assumed that most of the technology development would be done under contract. Contract administration and technical monitoring costs are NOT included in the projections, which are in 1982 dollars.

(c) III-V laser components are not included in the study.

Please note that programs that have been judged adequately covered by present NASA-sponsored effort are not included in the recommendations. For example, ongoing and presently planned efforts in the area of switches and switch matrices will likely result in components adequate for future space systems.

B. TARGET PROGRAMS

To focus more precisely on the recommended technology-development programs, classes of applications and their component needs were examined and tabulated. The results are presented in table VII-1 and discussed in some detail below.

TABLE VII-1. TARGET PROGRAMS

Program	Payoff	Applications
1. High-reliability solid-state modules	Multielement antennas	Large structures
2. Electronically adjustable components	Reconfigurable transponders	Future C-band systems
3. Ternary-compound amplifiers	"Superlinearity"	Spread-spectrum systems; Q-A modulation
4. Gigabit logic components	Digital processing	On-board switching; large structures; digital systems
5. Millimeter-wave components	Millimeter-wave transponders	Intersatellite links
6. L-band miniature hybrid transponder	High-efficiency, ultralightweight components	Personal communications ("wrist-watch transponder")

1. Large Structures

Large structures in space will definitely be built in the future, and multielement transmit/receive antennas are likely to form the heart of future space communications systems. Whether fixed-beam, switched-beam, or multibeam antenna systems are employed, high-reliability amplifier modules (containing switch elements, phase shifters, amplitude levelers or controllers, etc.) will constitute the key components of such antenna systems. For more accurate control of the antenna patterns, both amplitude and phase control of clusters of modules will be required, and the development of "constant efficiency" components - i.e., units in which efficiency does not decrease drastically as the level of operation is changed - will be a most desirable technical goal. If solar panels integral with the antenna elements are used (RCA's so-called SMART concept), the module designs will have to be capable of accommodating the solar cells, and the thermal dissipation problems solved accordingly.

2. Electronically Adjustable Components

One of the ways in which the operation of spaceborne communications transponders can be improved is to make them capable of changing their transmission characteristics on command from an Earth-bound controller. At present, transponders are prearranged to transmit, e.g., voice or video channels, or to operate in a single-carrier (in saturation) or multicarrier (linear) mode. If this arrangement can be changed, such "reconfigurable" transponders may be particularly useful when the spectral slots for present C-band systems are filled and there is a need to replace existing satellites with newer and more effective units.

The basic technology for developing the components of such a "reconfigurable" transponder is strongly based on III-V compounds. Filters of changeable bandwidth, switches, level adjusters, etc. are likely to use gallium arsenide as the basic semiconductor material. In monolithic, miniature-hybrid, or discrete form, components of this type will become powerful tools for the designer of future communications systems.

3. Ternary-Compound Amplifiers

Present investigations indicate that amplifiers based on ternary-compound MISFETs may exhibit better linearity than the more familiar GaAs MESFETs. "Superlinear" amplifiers may be of crucial importance in future systems using more-advanced modulation schemes or some form of spread-spectrum approach. It is thus important to continue the technology development of ternary-compound devices and associated microwave circuits.

4. Gigabit-Rate Components

Digital systems of the future will require IC components operating at gigabit rates or faster. Some of these components will have to be placed in the spacecraft, with attendant requirements of reliability, redundancy, etc. GaAs components now under development are logical candidates for future space communications service.

As discussed in Section VI of this report, enhancement-mode logic components have the advantage of lower power dissipation (a small fraction of that of depletion-type devices), better use of GaAs real estate, and simpler circuits. Yet, enhancement-type logic circuits are currently being pursued at a very low level in the United States. In Japan, on the other hand, such monolithic circuits have been demonstrated to perform at speeds comparable with those measured with depletion-type devices in the U.S. It seems important to pursue enhancement-type-logic monolithic circuits for future digital communications systems operating in space, independent of the carrier frequency employed.

5. Millimeter-Wave Components

Intersatellite links of the future, through which spaceborne transponders can communicate without interference and without the intervention of Earth stations, will probably employ carrier frequencies of 60 GHz or above. At present the only viable active components for such service are two-terminal devices (IMPATTs), usually based on the silicon technology. Three-terminal devices (FETs), if available, would enhance operation, simplify circuit design, and increase the reliability of millimeter-wave components. The upper frequency limits of III-V compound FETs produced by advanced and novel fabrication techniques have not yet been established. It appears important to pursue technology-development efforts aimed at developing FET-type devices for the millimeter-wave range.

6. Personal Communications/Mobile Systems

Future mobile services will permit communications between any two vehicles via satellite link. The ultimate development in this type of communications system will be the personal transponder, the size of a wrist watch, through which individuals will be able to communicate with each other by accessing a satellite equipped with multiple-beam (or switchable-beam) capabilities for regional or countrywide coverage. Extremely lightweight, highly efficient transponders operating in L-band will be required for such service.

C. TECHNOLOGY-DEVELOPMENT PROGRAM RECOMMENDATIONS

The following pages contain recommended technology-development programs for future space communications systems. For ease of reference, the proposed program descriptions follow a common format and appear on separate pages. They are numbered consecutively, with alternatives indicated by letters. There is no order of preference - technical or programmatic - implied in the numbering of the programs. Parallel programs for the same technology development are not included in the indicated approximate costs.

1(a) - ANTENNA MODULE TECHNOLOGY (MONOLITHIC)

Objective: Demonstration of a module suitable for use in a multielement antenna and containing, as a minimum:

- a radiating element
- an amplifier
- a phase shifter
- an amplitude controller

Frequency: ∿20 GHz

Power Level: 1 W, 6-dB gain

Timing: 3 years

Estimated Cost: $1.2M

Technology: Monolithic GaAs

1(b) - ANTENNA MODULE TECHNOLOGY (MINIATURE HYBRID)

Objective: Demonstration of a module suitable for use in a multielement antenna and containing, as a minimum:

- a radiating element
- an amplifier
- a phase shifter
- an amplitude controller

Frequency: ∿20 GHz

Output Power: 0.2 W

Gain: 13 dB

Timing: 2 years

Estimated Cost: $800K

Technology: Miniature hybrid with GaAs-FET active elements

1(c) - ANTENNA MODULE TECHNOLOGY (SMART)

Objective: Demonstrate cluster of antenna modules [similar to 1(a) or 1(b)] having radiating elements on one side of the flat structure, and solar-cell panels sufficient to power the cluster on the other side. Fabricate subarray with four amplifiers (16 antenna elements).

Frequency: 12 GHz

Output Power: 0.1 W per 4 antenna elements

Timing: 2 years

Estimated Cost: $600K

Technology: Miniature hybrid, GaAs FETs

2(a) - ADJUSTABLE TRANSPONDER COMPONENTS (SWITCHED)

Objective: To develop and demonstrate transponder components that could be adjusted, by switching commands from Earth, to provide transmission of different bandwidth and linearity characteristics.

Transponder Characteristics:

Frequency: 3.7-4.2 GHz

Output Power: 8.5 W

Gain: 55 dB

Timing: 2 years

Estimated Cost: $400K

Technology: GaAs FETs

2(b) - ADJUSTABLE TRANSPONDER COMPONENTS (ELECTRIC)

Objective: Same as 2(a) - except that all-electronic controls (rather than switching in and out of stages, etc.) will be employed.

Timing: 3 years

Estimated Cost: $750K

Technology: GaAs FETs

3 - TERNARY-COMPOUND LINEAR AMPLIFIERS

Objective: Investigate GaInAs MISFETs to determine their applicability as "superlinear" amplifiers for transponder service. Build demonstration amplifier for such an application.

Frequency: 3.7-4.2 GHz

Output Power: 8-10 W

Gain: ~50 dB

Timing: 5 years

Estimated Cost: $2.5M

Technology: GaInAs MISFETs

4 - HIGH-SPEED DIGITAL CIRCUIT TECHNOLOGY

Objective: Demonstrate LSI circuits such as multiplexers, memories, programmable dividers, etc., suitable for space communications applications. Develop fabrication processes and packaging of such circuits.

Timing: 5 years

Estimated Cost: $5M

Technology: GaAs FETs (enhancement mode)
GaInAs MISFETs (enhancement mode)
GaInAs-InP (heterojunction bipolars)
GaAs-GaAlAs (heterojunction bipolars)

5(a) - MILLIMETER-WAVE DEVICE INVESTIGATIONS (TERNARY COMPOUNDS)

Objective: Establish feasibility of three-terminal devices operating at millimeter wavelengths.

Frequency: 60 GHz and above

Timing: 3 years

Estimated Cost: $750K

Technology: GaInAs FETs

5(b) - MILLIMETER-WAVE DEVICE INVESTIGATIONS (PERMEABLE-BASE TRANSISTOR)

Objective: Same as 5(a).

Frequency: 60 GHz and above

Timing: 4 years

Estimated Cost: $1M

Technology: Permeable-base transistor (vertical FET), GaAs

5(c) - MILLIMETER-WAVE DEVICE INVESTIGATIONS (QUASI-BALLISTIC)

Objective: Same as 5(a).

Frequency: 60 GHz and above

Timing: 5 years

Estimated Cost: $1M

Technology: Planar-doped GaAs

6 - HIGH-EFFICIENCY, LIGHTWEIGHT, MINIATURE GROUND TRANSPONDER

Objective: Develop very small (wrist-watch type) transponder module for personal ground-to-satellite communications system. Emphasis is on a very high efficiency, ultralightweight amplifier and receiver technology.

Frequency: L-band

Power Output: 1 W

Efficiency: 70%

Timing: 2 years

Estimated Cost: $500K

Technology: Miniature hybrid, with GaAs FETs

VIII

Concluding Remarks

Future space communications systems are likely to show technical advances that, for purposes of analysis, can be grouped into three categories: (1) extensions of present systems; (2) systems in which new types of modulation will be employed; and (3) novel system concepts based on progress in components technologies. The study described in this report examined microwave components based on III-V compounds that require development to fill these future system requirements.

Large antenna structures employed in future systems will require quantities of antenna modules which - depending on the frequency of operation and antenna characteristics - could be fabricated in either monolithic or miniature-hybrid form. To achieve better utilization of the crowded spectrum, "superlinear" amplifiers will be likely to replace units in the currently used frequency region. The concept of a transponder with characteristics (bandwidth, linearity) changeable on command from Earth offers a flexibility not obtainable with present-day systems. Intersatellite links of the future may well operate at millimeter wavelengths, possibly doing away with primary power distribution problems by the use of solar cells to power amplifier modules directly. Digital systems will require high-speed monolithic ICs.

Advances in III-V materials technology (molecular-beam epitaxy, organometallic epitaxy, the use of vapor-phase epitaxy for growing ternary and quaternary compounds) have made new transistor geometries possible. These show promise of outperforming conventional planar GaAs FETs at microwave and millimeter-wave frequencies. This may be a particularly fruitful area of research.

The conclusions of the study are contained in a number of specific recommendations for research programs aimed at components to be employed in the identified applications.

Other Noyes Publications

ELECTRODEPOSITION PROCESSES, EQUIPMENT AND COMPOSITIONS

Edited by J.I. Duffy

Chemical Technology Review No. 206

This book covers over 200 recent developments in electrodeposition processes, equipment and compositions. The biggest advance in the art has been the increasing use of epoxy resin coatings. The epoxy coatings, obtained by cathodic deposition techniques, exhibit excellent resistance to corrosion, chemicals, and physical abuse, and they have above-average insulation properties. Other topics covered are anodic deposition, bath additives, curing agents, and amine-functional coatings.

Electrodeposition of water-based coatings, also termed "electrocoating," is a widely-used process with many advantages over other coating methods such as spraying, dipping, and rolling. Electrodeposition processes deposit a film of uniform thickness on essentially any conductive surface, even those with sharp points and edges. The films, when applied, are relatively water free and, thus, won't run or drip when removed from the bath. Little or no organic solvent is used in the resin system being deposited, therefore the processes are essentially fumeless and don't require extensive fume collection or incineration equipment. Additionally, a second or top coat can be applied over the electrodeposited film and both can be cured in one operation, thus effecting considerable cost reductions. Electrodeposition achieves widest use in the automobile and toy manufacturing industries.

The condensed table of contents listed below gives **chapter titles and selected subtitles.** Parenthetic numbers indicate the number of processes per topic.

ISBN 0-8155-0898-0 (1982) **308 pages**

Other Noyes Publications

DEPOSITION TECHNOLOGIES FOR FILMS AND COATINGS
Developments and Applications

by

Rointan F. Bunshah

John M. Blocher, Jr.
Thomas D. Bonifield
John G. Fish
P.B. Ghate
Birgit E. Jacobson
Donald M. Mattox
Gary E. McGuire
Morton Schwartz
John A. Thornton
Robert C. Tucker, Jr.

This book presents a unique collection of current knowledge on deposition technologies up to and including the most recent state of the art. It breaks new ground in the extensive coverage for those processes used in high technology, covering the entire spectrum from thin films to bulk coatings.

The contents deal with various technologies for the deposition of films and coatings, and the resulting microstructure, properties and applications. It covers the subjects of Evaporation, Ion Plating, Sputtering, Chemical Vapor Deposition, Electrodeposition from aqueous solution, Plasma and Detonation Gun Coating techniques and Polymeric Coatings. Additionally several other subjects common to the above technologies are also covered. They include: Adhesion of Coatings, Cleaning of Substrates, role of Plasmas in Deposition Processes, Characterization of Thin Films, and the Application of Deposition Technologies in Microelectronics.

Written by leading researchers in the field, this book will be useful and needed by scientists, engineers and managers working in industries associated with coatings for optical, electrical, mechanical, chemical and decorative applications. A **complete listing of chapter titles** is given below.

CONTENTS

ISBN 0-8155-0906-5 (1982) **585 pages**

Other Noyes Publications

ION IMPLANTATION FOR MATERIALS PROCESSING

Edited by

F. A. Smidt

Naval Research Laboratory
Washington DC

With an Introduction by E. Gelerinter and N. Spielberg
Kent State University

Chemical Technology Review No. 224

This book reviews current research on ion implantation for materials processing. The objective of the research described is to establish ion implantation as a viable technique for improving the surface properties of metals and alloys, those properties of particular interest being wear, fatigue, and corrosion. An introductory section on new potential applications of ion beam technology, of which ion implantation is one facet, has been included and provides excellent background information.

While ion implantation has been used extensively in the electronics industry for doping semiconductors, it may have even greater potential for the surface treatment of metals. Industry spends billions each year fighting wear and corrosion, the two leading killers of metals. The treatment of metal surfaces with lasers and ion beams shows great promise for providing improved corrosion resistance as well as greater hardness and strength. The advantage of ion implantation lies in its ability to produce a graded alloy, from surface to underlying bulk, such that both surface and bulk can be independently optimized.

The book is presented in four parts which cover ion implantation science and technology, wear and fatigue, corrosion, and other exploratory areas. Both fundamental and applications-oriented research are described in order to provide an understanding of the physical and metallurgical changes effected in the implanted region and to demonstrate the benefits of ion implantation. Applications studied include titanium implantation to extend machine tool life, carbon implantation to cut the effect of fretting on fatigue life, and titanium implantation in steel to increase corrosion resistance.

ISBN 0-8155-0961-8 (1983) **244 pages**